Condition-based Maintenance and Machine Diagnostics

Condition-based Maintenance and Machine Diagnostics

John H. Williams, *Reader*

Alan Davies, *Lecturer*

and

Paul R. Drake, *Lecturer*

at the Cardiff School of Engineering,
University of Wales,
College of Cardiff

CHAPMAN & HALL

London · Glasgow · Weinheim · New York · Tokyo · Melbourne · Madras

Published by Chapman & Hall, 2–6 Boundary Row, London SE1 8HN, UK

Chapman & Hall, 2–6 Boundary Row, London SE1 8HN, UK

Blackie Academic & Professional, Wester Cleddens Road, Bishopbriggs, Glasgow G64 2NZ, UK

Chapman & Hall GmbH, Pappelallee 3, 69469 Weinheim, Germany

Chapman & Hall Inc., One Penn Plaza, 41st Floor, New York NY 10119, USA

Chapman & Hall Japan, Thomson Publishing Japan, Hirakawacho Nemoto Building, 6F, 1-7-11 Hirakawa-cho, Chiyoda-ku, Tokyo 102, Japan

Chapman & Hall Australia, Thomas Nelson Australia, 102 Dodds Street, South Melbourne, Victoria 3205, Australia

Chapman & Hall India, R. Seshadri, 32 Second Main Road, CIT East, Madras 600 035, India

First edition 1994

Typeset in 10/12pt Times by Thomson Press (India) Ltd, New Delhi

Printed in Great Britain by St Edmundsbury Press, Bury St Edmunds, Suffolk

ISBN 0 412 46500 0

A catalogue record for this book is available from the British Library

Library of Congress Catalog Card Number: 94-70235

∞ Printed on permanent acid-free text paper, manufactured in accordance withANSI/NISO Z39.48–1992 (Permanence of Paper).

To Thelma, Sylvia and Rhian

Contents

Preface

A high level of machine availability is essential in flexible manufacturing systems. Unplanned stoppages are a major problem for machine maintenance. This text considers the use of condition monitoring as a means of applying predictive methods to manufacturing facilities in a planned maintenance policy where prevention is better than cure.

Successful maintenance depends upon a sound management structure together with robust diagnostic techniques. Both these areas are addressed in detail which will enable not only the undergraduate and postgraduate student to be well versed in the underlying concepts but also the management of manufacturing plants to implement sound maintenance practice.

We have diverse industrial and research experience which, we hope, has enabled us to marry the conceptual and practical constraints to present good practice in efficient maintenance. It is frustrating to see cars valued under £10 000 having in-built monitoring as a routine feature whilst machine tools worth over a quarter of a million pounds have none. Condition monitoring should be an integral part of any manufacturing facility design and not an afterthought.

John H. Williams
Alan Davies
Paul R. Drake
Cardiff, 1994

Introduction – aspects of maintenance

1

1.1 INTRODUCTION

Maintenance is a word with which most people are familiar. It is not essential to be an engineer to appreciate that machines fail, and that their failure may be delayed or even avoided by regular maintenance. Our current civilization for example, especially in the world's advanced nations, has been built up largely around the application of machines of one sort or another to virtually all aspects of everyday life. Thus a lay person readily understands the concept of maintenance and the consequences which are likely to occur, if it is not conducted correctly, and on a regular basis, for machines in their care or ownership.

A term which may however be unfamiliar to many engineers, let alone lay people, is 'Terotechnology', and yet this is the basic philosophy to which a majority of people adhere, when considering the purchase, care and maintenance of assets in their possession. The formal definition of Terotechnology is (BS 3811:1984):

> A combination of management, financial, engineering, building and other practices applied to physical assets in pursuit of economic life cycle costs.

This definition is expanded in the reference to note that in practice, Terotechnology is concerned with the specification and design for reliability and maintainability of plant, machinery, equipment, buildings and structures, with their installation, commissioning, operation, maintenance, modification and replacement, and with feedback of information on design, performance and costs. When translated for the lay person, Terotechnology can thus be simplified to the achievement of the best value for money via techniques which are many and various in their formality, approach and application (Hodges, 1991).

In this book the policy of condition-based maintenance (CBM) is advocated within an industrial context, that is to say its application to manufacturing plant and equipment as part of the overall philosophy of Terotechnology. The approach requires the use, and analysis of, actual condition data and accordingly it is important to recognize that the techniques outlined in the text are heavily dependent upon computer implementation.

Condition-based maintenance is one of a number of systematic approaches which can be used within the overall strategy of planned maintenance. Indeed the optimized operation of a maintenance department within a manufacturing company may dictate the use of several of these approaches in combination. Accordingly, it is worthwhile placing CBM in its proper context, by first reviewing what we mean by maintenance in a manufacturing sense and then by outlining the various maintenance approaches available. Maintenance has been defined as (BS 3811:1984):

> The combination of all technical and associated administrative actions intended to retain an item in, or restore it to, a state in which it can perform its required function. This required function may be defined as a stated condition.

The key words to note in this definition identify the fact that there are both technical and administrative actions involved in maintaining an item of equipment in a satisfactory and functional state. Thus maintenance actions can be split into two distinct categories, that is those without a logical and predetermined administrative plan, and those organized with forethought to produce a logical and predetermined administrative plan of action.

The second group contains many different approaches, one of which, as shown in Fig. 1.1 is condition-based maintenance. Each category has its strengths and weaknesses, accordingly both will be reviewed, so that the techniques of condition-based maintenance can be understood within their proper context. Before the review is undertaken however, it is essential to explain why maintenance is now being recognized as an important activity within a manufacturing concern, and one which significantly impacts the competitiveness and profitability of a company.

The introduction of highly automated equipment such as flexible manufacturing systems or cells (FMS/FMC) to the discrete batch manufacturing sector of UK industry, has produced a realization that downtime, and its associated lost production, is a great deal higher and more costly on this type of equipment than with the more traditional, non-integrated manufacturing plant which it replaced. Indeed when one considers the cost

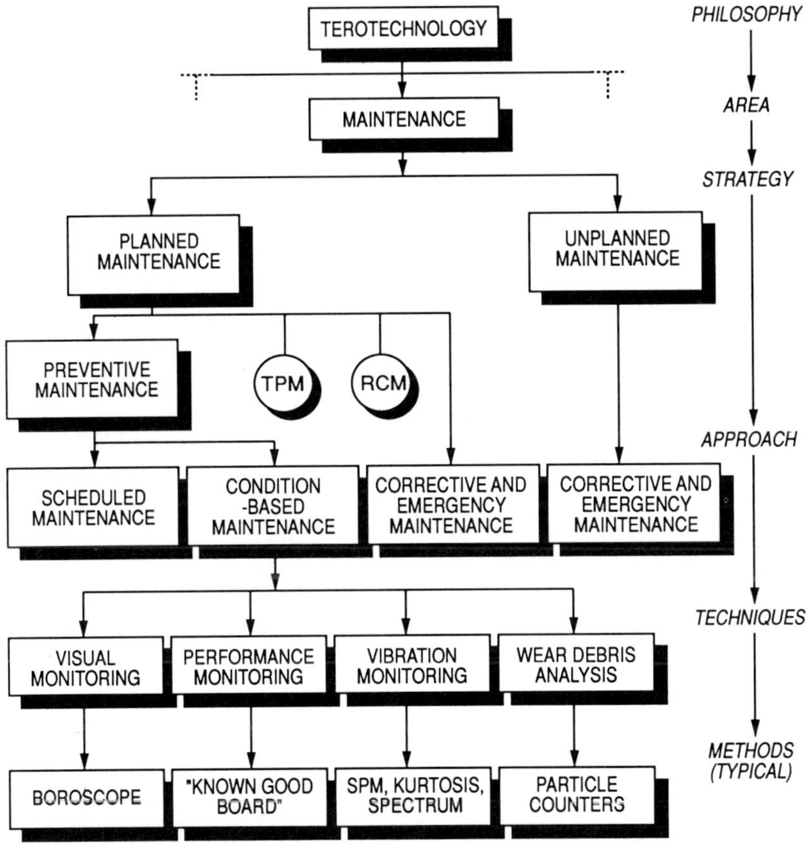

Fig. 1.1 Relationship of CBM to various forms of maintenance. TPM = Total Productive Maintenance. RCM = Reliability Centred Maintenance.

involved in the failure and repair of this type of hi-tech equipment, it is easy to understand why system availability, reliability and maintainability are now regarded as being of first importance to manufacturing managers.

Hence it is essential to understand the relationship between these three parameters and to note that high plant availability is absolutely dependent upon the inherent reliability and maintainability incorporated in an equipment's design.

The initial value of a system's availability is thus governed by the size of each of the two parameters above, and further it is important to note that these values are decided during the equipment's functional design. It is also important to remember that subsequent plant availability is dependent upon maintaining these values, which over time will tend to degrade with normal equipment usage.

Plant availability in a manufacturing context can therefore be defined as:

The time the system is available to do useful work.

So mathematically, the steady state equipment availability can be expressed as follows for continuously operable and maintained systems:

$$\text{Availability} = (\text{MTBF}/(\text{MTBF} + \text{MTTR})) \times 100\% \qquad (1.1)$$

Note that MTBF is the mean time between equipment failures, and that MTTR is the mean time to repair the equipment after failure. The above equation, within reliability theory, is often referred to as the 'uptime ratio' or 'time availability' in order to distinguish it from any other definition of availability which might be more applicable under certain circumstances (Calabro, 1962; Lie *et al.*, 1977; Nakajima, 1984).

It is obvious from the relationship that an increase in MTBF, or a decrease in MTTR, will result in an increase in system availability, as the MTBF is related to the reliability of the system through its failure rate (λ), i.e. how often it fails, and the MTTR is linked to its maintainability via the maintenance action rate (μ), i.e. how easy it is to repair. By examination of equation 1.1 both can be seen to be crucial to the performance of the system in terms of its availability for production.

System reliability and maintainability are, as previously stated, an inherent part of the functional design of an item of equipment, therefore improving their value retrospectively can be quite a difficult, costly and complex task. Thus the importance of considering both factors during system specification and design can not be overemphasized. Indeed design for functionality must be complemented with design, if possible, for failure free operation or failing that, minimal and easy maintenance. This is particularly true in the case of complex and highly automated manufacturing plant, which by their very nature, are likely to contain extremely valuable and critical machines within an integrated production system.

Maintenance should thus be more than simply an activity to repair failed equipment, for by that definition no advantage is gained by the system user in operating a maintenance department. Indeed the user loses by virtue of:

1. A probable decrease in MTBF (no system is quite as good as it was new after it has been repaired). So system reliability has declined and failure frequency increased. The quality of product output may also be affected.
2. A possible increase in MTTR due to an increased failure frequency. This can result in a high spares usage with its associated costs, and opportunities for logistic delays in the spares procurement/equipment repair process.

3. Production losses in terms of product output, the associated costs and quite possibly costs linked to missed delivery times.
4. The loss of future sales due to customer dissatisfaction. Not readily quantifiable but a distinct possibility in many cases.

It is against this background therefore, that we need to examine firstly the alternative strategies in maintenance, their advantages/disadvantages, associated approaches, and later in the text the specific techniques/methods involved in condition-based maintenance.

1.2 UNPLANNED MAINTENANCE – NON-SYSTEMATIC PROCEDURES

This strategy is often known more colloquially as 'run to failure', essentially it is the 'do nothing until it breaks' option. Formally it may be defined as follows (BS 3811:1984):

> The maintenance carried out to no predetermined plan.

or more informally as, the equipment continues in operation until it fails, at which point it is either repaired or replaced (Davies, 1990).

Two situations can be visualized under this strategy in respect of production plant or equipment, their differentiation being dependent upon the interpretation of a failure and its contextual circumstances.

1.2.1 Corrective maintenance

This on the surface appears to be a straightforward case, viz. there is a failure of the equipment which brings production to a halt, and which requires the equipment to be repaired before production can restart. In essence corrective maintenance is (BS 3811:1984):

> The maintenance carried out after a failure has occurred and is intended to restore an item to a state in which it can perform its required function.

The bare definition however needs some additional qualification, so that when interpreted, it is understood to include a dependency upon the type of failure and its circumstances to mean, the immediate or delayed repair of a non-serious, in terms of production or safety, catastrophic or degradation failure.

In practice this revision now covers the case where simple low cost systems fail, or move out of tolerance, and which are duplicated within the manufacturing system, so that it is easy to shift and maintain production on a second machine whilst the first is being repaired. By definition it also covers the case of a single machine which on failure is repaired. This sort of hairsplitting is important, because:

1. It illustrates that within what appears to be a rigid strategy there is some flexibility, endowed by the interpretation of a failure and its manufacturing context.
2. Such flexibility can be used to some effect in the costwise optimization of maintenance.

1.2.2 Emergency maintenance

This approach to maintenance is often confused with corrective maintenance, however its precise definition illustrates the difference between the two (BS 3811:1984):

> Emergency maintenance is the maintenance which it is necessary to put in hand immediately to avoid serious consequences.

Note that in this case, constraints are applied which limit the flexibility of maintenance action and thus its possible costwise optimization. These constraints are expressed in the definition as the immediacy of the required action and the implication that serious consequences will follow if it is not carried out. The qualification previously applied to the definition of corrective maintenance, in respect of the type of failure and its contextual circumstances is also valid here. However it should now be restated as, the immediate repair of a serious, in terms of production or safety, catastrophic or degradation failure.

The practical implications of a lack of flexibility and cost control in this category of maintenance therefore, makes it a highly undesirable approach to adopt. Consequently, the requirement for emergency maintenance should if possible be minimized within any maintenance organization by the adoption of more flexible cost effective policies.

Unplanned or breakdown maintenance is therefore suitable only in circumstances where unscheduled stoppages cause minimal inconvenience, and its advantage lies in the fact that under these conditions, it may well be a low cost option. Indeed where components fail randomly and provide no prior indication of impending failure, this type of maintenance remains the only option. In the case of critical manufacturing equipment however, unscheduled stoppages cause great inconvenience and in an integrated production system random machinery failure can result in expensive scrapped components together with punitive maintenance and lost output costs.

1.3 PLANNED MAINTENANCE – SYSTEMATIC PROCEDURES

In contrast to the previous strategy, planned maintenance is (BS 3811:1984):

> The maintenance organized and carried out with forethought, control and the use of records to a predetermined plan.

Encapsulated within this formal definition is the implied management of the maintenance function; within the case of manufacturing, the objective of achieving high production system availability and excellent maintenance system efficiency at minimum cost. Note that preventive maintenance is always a part of planned maintenance, whilst corrective or emergency maintenance as we have already seen, may or may not be.

The kaizen or continuous improvement aspect of total productive maintenance (TPM) for example, can be categorized as being planned corrective maintenance while the overall viewpoint of TPM is that of planned preventive maintenance. A knowledge of the distinctions between the subgroups or approaches to planned maintenance is therefore essential to achieve good maintenance management, for they are not mutually exclusive, and can be employed on a 'mix and match' basis in the optimization of a company's maintenance operation (Wilmer, 1992).

1.3.1 Preventive maintenance

The formal definition of preventive maintenance is (BS 3811:1984):

> The maintenance carried out at predetermined intervals or corresponding to prescribed criteria and intended to reduce the probability of failure or the performance degradation of an item.

Embodied within this definition is the desire to in some way prolong the effective operation, availability or useful life of a system, by conducting the regular inspection of, and/or maintenance on, the equipment in question.

Obviously the trick is to do so cost-effectively, given that the maintenance department is a non value adding division of the company, and usually has severely restricted resources, both in manpower and money. Blanket short interval preventive maintenance on all production machinery within a company is therefore not a viable option because:

1. The manpower and money simply would not be available.
2. The use of preventive maintenance on some types of equipment can be inappropriate or ineffective, and result in expenditure without notice-

able improvement in performance or reliability. Indeed a degradation in both may occur due to maintenance induced failures.

It is important therefore to attempt to optimize our use of preventive maintenance, in a cost-effective sense, by identifying:

1. The appropriate critical production machinery on which it can be effectively used.
2. The most efficient method of determining the time interval or criterion which dictates when it should be undertaken.

This last requirement results in the subdivision of preventive maintenance, as outlined in Fig. 1.1 to give the two traditional approaches which are scheduled maintenance and condition-based maintenance. However in recent years two additional approaches have found favour in some sectors of manufacturing industry, the newcomers being reliability centred maintenance (RCM) and total productive maintenance (TPM). Note that TPM attempts to offset the possibility of maintenance induced failures by extensive, and expensive staff training (Nakajima, 1984).

The relationship of these four approaches to preventive maintenance and each other is shown in Fig. 1.2 which also outlines where opportunistic preventive maintenance can be used. This last named approach may be carried out on a machine whilst a breakdown is being repaired, and to some extent can offset the cost involved, by extending the system's availability prior to the next failure. A fine judgement is required regarding the use of opportunistic preventive maintenance for it to be cost effective as:

1. It may extend the original failure/repair downtime beyond the point of economic payback in terms of extended equipment availability.
2. It may induce failure if the relevant preventive maintenance procedure is not followed correctly.

1.3.2 Scheduled maintenance

This approach to preventive maintenance attempts to forestall the occurrence of breakdown by identifying on an historical basis, the duration of the failure interval exhibited by a component or machine. Accurate failure data in the form of maintenance records must therefore be available to establish typical component or machine failure patterns. An estimate can then be made of the relevant failure rate and appropriate maintenance interval. Thus the definition of scheduled maintenance is (BS 3811:1984):

> The preventive maintenance carried out to a predetermined interval of time, number of operations, mileage etc.

This approach has some inherent difficulties, mainly concerning the

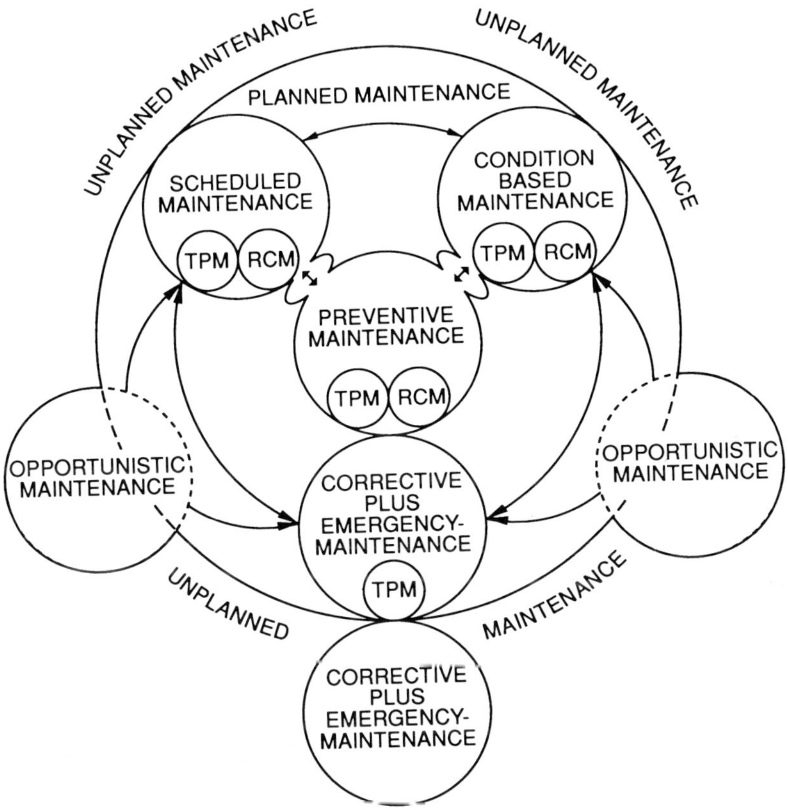

Fig. 1.2 Relationship of approaches within planned maintenance.

derivation of proven criteria whereby the predetermined time interval is established. The difficulties/disadvantages are outlined below, and need to be carefully considered before the implementation of a scheduled maintenance scheme is undertaken.

1. In cases where the estimate of failure time is conservative, a risk exists that the system or component may be replaced under the scheduled maintenance regime well before its useful life has elapsed. This results in the conduct of excessive and unnecessary maintenance, leading to the expensive overstocking of spare parts and the incurring of superfluous labour costs.
2. Conversely, if the estimate of failure time is optimistic, a risk exists that the system or component may fail in service, with all that this implies for the attendant inconvenience and subsequent cost.
3. When estimating the failure time, due consideration should be given to

any variation in the loading that a component or system may experience during its normal operation. Allowing for such variations can considerably complicate the analysis and the subsequent prediction of a probable failure time. This leads to wide variations in the accuracy of failure prediction and hence to the problems previously outlined.

4. As already stated, great care should be used when undertaking scheduled maintenance, for when applied to all but the most simple of systems, the stripping down and refurbishment of equipment can of itself induce further system failures.

It should be noted that for scheduled maintenance, the predicted failure time is usually estimated initially via the application of reliability theory to 'a prior' field failure data for the system or component. If no prior data exists, the prediction can be based on 'similar' data, sometimes modified to include a best estimate of how the system or component in question will behave when in service.

Such predictions are often updated as actual field failure data becomes available from the equipment or item in question once it commences operation. The maintenance interval, which is based on the failure prediction, can also be adjusted in the light of experience, to suit the requirements of a manufacturing schedule such that the plant can be maintained when not in production.

1.3.3 Condition-based maintenance

This approach to preventive maintenance attempts to overcome the 'crystal ball' data analysis/prediction requirement inherent in scheduled maintenance. It does so by seeking to determine the actual operating condition of a system or component at any point in time. The establishment of the item's correct status, or condition at the required time, is obviously critical to the viability of the approach and to this end various monitoring techniques are used within CBM.

Formally condition-based maintenance is defined as (BS 3811:1984):

> The preventive maintenance initiated as a result of knowledge of the condition of an item from routine or continuous monitoring.

Or more informally as, components are periodically inspected by manual or automatic systems in order that their condition may be assessed and to identify their degradation rates. A decision is then taken regarding replacement and this is based upon an analysis of the monitored data (Seddon, 1984).

This concept, of undertaking maintenance only when it is required, and based on the actual condition of a system or component is obviously very attractive. Some of the perceived advantages are (Davies, 1990):

1. An early knowledge of impending failure, through the use of trend analysis techniques and/or qualitative inspection. Preventive maintenance put in hand as a consequence of this information reduces the system/component failure rate and so increases equipment reliability.
2. Actual condition knowledge and accurate failure prediction. This allows sufficient flexibility for the plant to be shutdown for repairs at a point in time convenient to the manufacturing schedule thereby improving system availability and minimizing random failure.
3. Diagnostic potential. This is present in a condition-based maintenance approach and can pinpoint impending component failure together with the cause of that failure. The use of diagnostic routines also permits system modification, to remove identified failure modes and thereby improve inherent design reliability. In addition, through the rapid identification of an impending failure, system maintainability is improved together with workload planning and spare part stockholding.
4. Flexibility. If used in conjunction with an adaptive control system, the monitoring aspect of condition-based maintenance can extend the useful life of plant and equipment, by limiting the effect of a dynamic failure mechanism within the system. This reduces the likelihood of early failure and thus improves system reliability.

Hence the ability to forecast failure at a time sufficiently in advance of the event for action to be taken is the fundamental premise upon which the approach of condition-based maintenance is built. The key component in this approach is thus the need to achieve an accurate failure prediction, and therefore an effective method of monitoring the status of an item or system is essential. Clearly the ideal technique would be one in which the condition of the equipment would be known at all times and which would provide an accurate prediction of any potential failure on demand. Attempting to fulfil this requirement to a greater or lesser degree are the techniques embodied within condition monitoring (CM).

Formally, condition monitoring is defined as follows (BS 3811:1984):

> The continuous or periodic measurement and interpretation of data to indicate the condition of an item to determine the need for maintenance.

The standard also notes that condition monitoring is normally carried out with the item in operation, in an operable state or removed but not subject to major stripdown. More colloquially, condition monitoring has been described as the assessment of the current condition of plant and equipment by the use of techniques which can range from sophisticated computer-driven instrumentation to human sensing, in order to predict failure and to economically perform maintenance only when a potential failure is identified and at a time convenient to the production schedule (Davies, 1990).

Consequently, condition monitoring is the means by which condition-based maintenance can be carried out, utilizing as it does the various techniques available to monitor and identify impending failure. The use of condition monitoring thus ensures that all necessary maintenance actions can be undertaken at a time appropriate to the predicted failure, and therefore allows the maximization of equipment availability through the timely and systematic organization of maintenance. In essence there-

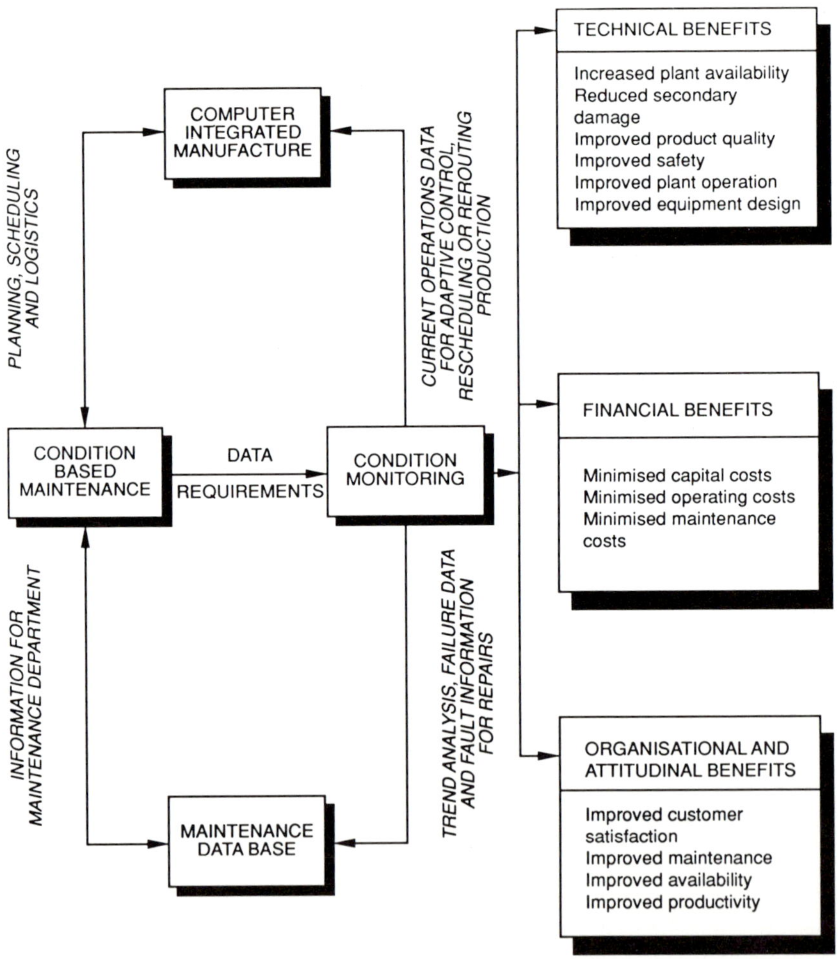

Fig. 1.3 Application of condition monitoring in computer-based maintenance.

fore the implementation of condition monitoring can be described as, the performance of periodic or continuous comparative measurement on parameters which are suspected of reflecting the condition of a component, sub-assembly or system with the object that on analysis, the measurements may indicate the item's current condition and the future trend of its possible deterioration (Davies, 1990).

Thus through its contribution towards the planning of future maintenance action, condition monitoring ensures that all decisions are made on substantive and corroborated diagnostic information, thereby providing a basis for cost-effective and logical decision making (Collacot, 1980). The integration of condition monitoring techniques within the wider approach of CBM can yield many advantages, some of which are outlined in Fig. 1.3 (Davies, 1990).

1.3.4 Reliability centred maintenance

The concept of reliability centred maintenance (RCM) originated in the airline industry, where the cost of safety-driven, failure preventive servicing programmes and poor aircraft availability prompted the need for a new approach to maintenance. Reliability centred maintenance has been defined as (Kumar and Granholm, 1990):

> A systematic approach to quantitatively assessing the need to perform or revise preventive maintenance tasks and plans.

It provides a tightly focused methodology which is targeted on system function, the failures relating to that function, and in particular to the effects of dominant functional system failures. A decision tree is used within RCM to identify and classify critical system components together with an appropriate and applicable maintenance policy. The main concept underlying the development of RCM is, an attempt to retain the designed reliability of equipment, through the analysis of factors which affect its operating reliability, and with a view to optimizing preventive maintenance programmes via effective maintenance planning (Kumar and Granholm, 1990).

Thus the main objective of RCM is the cost-effective maintenance of an equipment's inherent reliability value. It does so by utilizing the principal justification criteria of safety, availability and cost-effectiveness to decide what maintenance should be undertaken to achieve the deferral or prevention of specific failure modes (Moss, 1985).

Reliability centred maintenance is implemented via twin analysis routines investigating the system design. One analysis seeks to identify the origin of potential failure modes and their probability of occurrence, whilst the other develops the decision logic for each mode, and provides the formulation structure necessary to determine the nature of the scheduled maintenance tasks required. The two major aspects of RCM are therefore:

1. The determination of a maintenance requirement and the identification of the tasks necessary to satisfy it.
2. The reliability analysis of critical system components and hence the determination of the maintenance task interval.

Evaluation techniques such as failure mode effects and criticality analysis (FMECA) or fault tree analysis (FTA), are used to determine failures which may affect the operational capability of the equipment thereby disclosing the significant maintenance tasks. In addition, by partitioning the equipment into object categories, significant items can be identified, which through intensive decision theory (DT) analysis permit the determination of maintenance actions.

As with scheduled maintenance, the intervals between maintenance tasks are estimated, using 'a prior' statistical failure data and a knowledge of the appropriate probability distribution. In some cases however, actual condition knowledge obtained via condition monitoring, has been used to determine the physical state of critical items and hence permit a dynamic determination of the required maintenance interval (Kumar and Granholm, 1990; Sandtorv, 1991).

Reliability centred maintenance provides a useful cost control methodology whilst simultaneously maintaining the inherent reliability of the equipment. In addition, the analyses used in RCM provide a means by which any inherent equipment design failures may be located, whether scheduled maintenance is required, the time interval for such maintenance and what the maintenance tasks may be. Considerable cost savings are claimed for this approach to maintenance along with optimum repair effort, maximum safety and high productivity (Kumar and Granholm, 1990; Sandtorv, 1991).

1.3.5 Total productive maintenance

This approach to maintenance evolved in Japan, initially as a development of preventive maintenance, before being enhanced in several stages to finally become productive maintenance. The development stages include the incorporation and application of breakdown maintenance, reliability and maintainability theory, concern for economic efficiency in plant design and ultimately the introduction of a comprehensive maintenance system based on total employee participation. Total productive maintenance has been defined as (Nakajima, 1984):

> Productive maintenance carried out by all employees through small group activities.

It is also important to note what is left out of this definition, that is total productive maintenance is plant maintenance conducted by the user of the equipment, on a company wide basis, and includes all the available methods and personnel to effect a solution to maintenance problems within the user company.

In this respect TPM recognizes that different maintenance situations within a company may well require differing methods to effect a solution. Accordingly, it uses whatever method or technique works cost effectively in a given situation. These may differ from machine to machine or plant to plant, however so long as it works and is cost efficient it is total productive maintenance. Many of the techniques used in TPM are not new, as illustrated by its development cycle; what is new however is the Japanese culture, its penchant for total employee participation and the link with small group activities.

The key contribution made by TPM to maintenance theory is that it destroys the artificial barrier or demarcation line between the maintenance and production departments within a company. Ideally, the ‘you break it we fix it’ mentality is completely eliminated when TPM is implemented effectively within a firm. This leads to huge gains in productive efficiency and the recognition that organizational problems in maintenance far outweigh the technical difficulties in their ability to make a manufacturing system ineffective and non-profitable.

In addition, the continuous improvement aspect of TPM removes all complacency within an organization and replaces it with a strong sense of purpose, that is a striving to achieve a zero equipment failure situation and hence maximum equipment availability, productivity and production quality. The introduction of TPM also removes, over a number of years, any dependency on outside expertise in equipment maintenance. It does so via an integral implementation requirement of full, complete and continuing staff education and training.

A complete definition of total productive maintenance thus includes (Nakajima, 1984):

1. The aim of maximizing production equipment effectiveness in terms of economic efficiency and profitability.
2. The establishment of a thorough system of productive maintenance, which includes maintenance prevention, maintainability improvement and preventive maintenance for an equipment’s complete life cycle.
3. The implementation of productive maintenance in the company by all relevant departments.
4. The involvement of all company employee’s from top management to shop floor workers in the implementation of productive maintenance.
5. The promotion of productive maintenance through autonomous small group activities.

Central to the TPM approach is thus a desire to eliminate, in every case, the root cause of lost production and thereby improve equipment effectiveness. In this respect, equipment availability or operating rate is defined within TPM as (Nakajima, 1984):

$$\text{Availability} = \{\text{operating time/production time}\} \times 100\% \tag{1.2}$$

or more quantifiably as:

$$\text{Availability} = \{(\text{SPT} - \text{DT})/\text{SPT}\} \times 100\% \tag{1.3}$$

where SPT is scheduled production time and DT is equipment downtime. Note that (SPT – DT) = APT or available production time.

Included in the downtime figure are stoppage losses due to failures, scheduled downtime for equipment changeover, setup or adjustment. It does not include planned downtime, scheduled in the production plan for maintenance and/or other management activities, nor does it include losses due to poor process efficiency or any other minor stoppages.

The above equation does not therefore tell the whole story, for only downtime losses are used to give a measure of availability, whereas in practice other factors need to be considered to provide a more accurate measure of equipment effectiveness. Accordingly, six major areas of concern are identified and included in the TPM overall measure of equipment effectiveness. These are (Nakajima, 1984):

1. Manufacturing system downtime: due to machine component failures, which result in consequential equipment breakdown.
2. Manufacturing system downtime: as a consequence of setup and adjustment. Typically due to production changeover requirements, and involving the exchange of tools or dies.
3. Manufacturing delay: caused by idling or minor stoppages. Possibly linked to sensor malfunction or workpart jamming in feed chutes.
4. Manufacturing delay: involving cycle time errors, induced via machinery speed variations.
5. Manufacturing process defects: resulting in poor product quality and hence lost output.
6. Manufacturing losses: due to extended machinery commissioning and startup time, thus delaying the point at which stable production is achieved.

Hence, overall equipment effectiveness (OEE) is measured in TPM by:

$$\text{OEE} = \{\text{A} \times \text{PE} \times \text{QE}\} \times 100\% \tag{1.4}$$

where A = equipment availability, PE = production efficiency and QE = quality efficiency.

Nakajima outlines in further detail the precise calculations involved in measuring overall equipment effectiveness. Although as shown, due to the multiplier effect embodied in equation (1.4), to achieve a world class performance of around 85% then (Nakajima, 1984):

1. Equipment availability needs to be greater than 90%.
2. Production efficiency should be higher than 95%.
3. Quality efficiency should equal 99% as a minimum.

Total productive maintenance has thus a main aim of maximizing equipment effectiveness within a company, and to do so, it places the responsibility directly on the equipment user. In attempting to achieve its goal, TPM provides a tight focused approach for the equipment using company. It does so by identifying, solving and keeping to specific internal user company maintenance problems. All difficulties are overcome via an employee participative and reinforcing resolution cycle. In this context TPM can be thought of as the implementation of best maintenance practice.

It is important to note that TPM does not address the wider and more generic maintenance problem, which involves equipment supply, suppliers or supply systems, as do the concepts of terotechnology or logistics. However as an approach to maintenance, within these concepts, as shown in Figs 1.1 and 1.2, it is none the less effective for that.

1.4 CONCLUSION

The previous sections have attempted to outline in brief detail some of the complex issues and approaches involved in maintenance. A more comprehensive treatment would however be required to explain all the subtleties involved and that would be worthy of a volume in its own right. In this text the main issue is to provide a complete explanation of computer-based maintenance and to that end the subsequent chapters are dedicated.

1.5 REFERENCES

BS 3811:1984 British Standard Glossary of Maintenance Management Terms in Terotechnology, British Standards Institution (1984).

Calabro, S.R. (1962) *Reliability Principles and Practices*, McGraw-Hill, New York.

Collacot, R.A. (1980) Why condition monitoring – a review. *Terotechnica*, **1** (3), 173–178.

Davies, A. (1990) *Management Guide to Condition Monitoring in Manufacture*, IProdE, London.

Hodges, W. (1991) Terotechnology. *Maintenance*, **6** (4), 22–25.

Kumar, U. and Granholm, S. (1990) Reliability centred maintenance – a tool for higher profitability. *Maintenance*, **5** (3), 23–26.

Lie, C.H., Hwang, C.L. and Tillman, F.A. (1977) Availability of maintained systems: a state of the art survey. *AIIE Transactions*, **9** (3), 247–259.

Moss, M.A. (1985) *Designing for Minimal Maintenance Expense*, Marcel Dekker Inc., New York.

Nakajima, S. (1984) *Introduction to Total Productive Maintenance*, Productivity Press Inc., Cambridge, MA.

Sandtorv, H. (1991) RCM – Closing the loop between design, reliability and operational reliability. *Maintenance*, **6** (1), 13–21.

Seddon, G.N.D. (1984) Computer applications for condition based maintenance – data collection, trend monitoring, future prediction and system integration. *Maintenance Management International*, **4** (4), 229–236.

Wilmer, B. (1992) Total Productive Maintenance (TPM) at Nissan Motor Manufacturing (UK) Ltd.

Maintenance information systems

2

2.1 INTRODUCTION

The increasing use of highly sophisticated automation in UK manufacturing industry, has introduced at the plant level, a requirement for electronic and computerized information handling to allow the effective management of this type of production equipment. Automation, sometimes called modern or advanced manufacturing technology (AMT), has been described as (NEDO, 1985):

> Any substantially relevant and new technique, the adoption of which is likely to lead to changes within a firm in manufacturing practices, management systems and approaches to design and production engineering of the product.

These automated manufacturing systems, by their very nature, must use computerized information technology (IT) to:

1. control each individual item of production machinery,
2. provide management information on system performance,
3. ensure that the equipment is integrated, and performs as a 'system' or single entity when in operation.

Information technology can be defined in a general sense as (Zorkoczy, 1982):

> The way in which we collect, store, process and use information.

In the context of automated manufacturing therefore, information technology might be more accurately described as, the mechanism by which instructions concerning production may be accurately encoded, transmitted, received, stored, processed and used. This ability linked with 'real time' computer control, data processing and 'expert' decision making allows Information Technology to supply the 'pseudo-intelligence' required for automated manufacture.

On this basis, Information Technology is applied at each level of control within an automated manufacturing system, and in addition, to the external production, maintenance and management functions which support the equipment. Thus a 'factory-wide' information system or network is formed to control the day-to-day operations of a manufacturing plant.

Manufacturing systems software is therefore an important feature of production equipment design, for it is the vehicle through which Information Technology is implemented, and as such, can form a sizeable part of the capital outlay required for the purchase of automated production equipment. The requirement for reliability, in both software and hardware, is a key characteristic of any specification for advanced manufacturing technology. Indeed these systems are usually critical 'bottlenecks' within a production plant, and equipment failure can result in punitive lost production costs aggravated by a lack of standby production facilities.

Thus it is obvious that for automated manufacturing systems, maintenance has become an essential company function, to ensure the availability and continued operation of this type of production facility. Unfortunately, because it does not appear to directly contribute to output, and hence company profit, it is often seen by managers and accountants as an annoying liability rather than a useful company asset. Some managers fail to appreciate that an investment in maintenance skills, equipment, and the conduct of correct maintenance policies are a direct contribution to the 'bottom line' via high plant availability and excellent product quality.

In this way, the introduction of a sophisticated production plant controlled by Information Technology has increased the strain on many maintenance departments, and although it may not be so obvious, on the efficiency of the manufacturing organization itself. As new production systems tend towards full automation and embody the increasingly sophisticated techniques involved with electronic information processing, the efficient gathering and evaluation of data relating to the system, its operation and maintenance becomes an essential requirement for optimum management control (Gadd, 1990).

Accurate 'real time' knowledge relating to the equipment and its operation are therefore vital to optimize overall machine performance. This translates in terms of plant reliability and maintenance into a requirement for computerized maintenance information and analysis, to maximize repair efficiency, system utilization, productivity and quality of output.

2.2 ORGANIZATIONAL STRUCTURE

One obvious and essential aspect of any maintenance information system (MIS) is that it must efficiently service the organization which has purchased and installed it. To design a new MIS, or modify an existing system therefore, requires a thorough knowledge of the maintenance organization which it is intended to serve.

Within industry, there are many differing maintenance organizational structures in use, in the main they are not 'designed' *per se*, but have developed over time to suit the particular needs of an individual company. Accordingly, understanding a particular organization for the purpose of MIS design requires some consideration, and most importantly a starting point in the form of knowledge of the department's function or mission within the company.

Various functions, mission statements or objectives have been promulgated within the academic literature for maintenance departments typical are those suggested by Kelly (1991):

> The function of a maintenance department is being concerned with providing and controlling the reliability of the plant, whilst its objective is to achieve the agreed output level/operating pattern, at optimum maintenance resource cost, and within the constraints of safety and plant condition standards.

In addition, the nature and mix of work undertaken by a maintenance department, together with the resources, techniques and strategy it uses to satisfy requests for service, play an important part in the evolution and final shape of the maintenance organization. Thus in essence, maintenance management is the art of matching resources to tasks with the objective of achieving the ideals encapsulated in the mission statement via the maintenance organization.

Three main types of maintenance organization can be identified; they are (Kelly, 1991):

1. Traditional. Used extensively in medium to large-sized companies, this approach provides for plant or equipment oriented, shift and day maintenance crews, supported by centralized resources and contract labour. In this design, maintenance is conducted exclusively by such a support organization, possibly with some local to plant work areas, but without assistance from production personnel. It has the disadvantages of low resource utilization consequential to a variable workload, antipathy between the workforce and management, plus a potential for conflict between maintenance/production personnel.
2. Centralized. In this design, the maintenance organization is based

around areas of plant rather than complete systems. The shift crew report to an engineering staff who work closely with production personnel, while the main maintenance support resources remain tightly centralized and controlled by the maintenance management. This type of organization has the advantage of increasing utilization via good planning, workload smoothing and work measured incentive schemes. However it has the disadvantage of a workforce which remains polarized in respect of production personnel and its own management.

3. Ship System. This type of maintenance organization is based on semi-autonomous multitrade units, who are responsible for a particular plant within a profit centre, and report via a plant engineer to a plant manager and thence to the production manager. A chief engineer controls a centralized workshop, providing maintenance services, resource support, and supervision of major project work on the plant. The main advantages of this structure are claimed to be the precise delegation of objectives and responsibilities, the linkage of production/maintenance personnel and the sense of ownership generated in plant oriented terms. Additional advantages also cited in support of this organizational design are reduced management levels, an improved role for the shopfloor supervisor, the multiskilling/working pattern flexibility of staff and the increased use of contractors.

In practice, these three main designs are supplemented/modified by the individual requirements of a particular company. Thus any MIS software must be carefully written, or modified, to serve the exact requirements of the customer in question and his particular organization. The case study outlined in sections 2.5 and 2.6 illustrates how the maintenance organization may be defined using the structured analysis and design technique (SADT), and also how this information can be subsequently used for MIS design and implementation.

2.3 SYSTEM SPECIFICATION

From the foregoing sections, it should be obvious that the design of an efficient maintenance information system is a somewhat complex undertaking, requiring not only a thorough knowledge of the maintenance organization involved, but also a clear understanding of what is needed/provided by the implementation of the MIS. The definition of what is needed should encompass the function, mission statement and objectives of the maintenance department, along with the identification of the techniques to be used and the performance parameters necessary to determine if these ideals are being met.

Thus in essence there are two primary objectives (Beadle, 1989):

1. Clarifying what is actually needed in an MIS as opposed to computerizing the current maintenance department practice.
2. Identifying and establishing the relevant techniques/performance parameters, which can provide a comparison/quantitative measure of improvement, between a proposed MIS and an equivalent manual system.

Achieving these objectives is a task for a project team, operating within the company, and headed by a manager who is fully conversant with the needs of the maintenance department. This requirement, together with the mission statement, objectives, and a knowledge of the function/interaction of maintenance with other departments within the company, provide a foundation for the design of the MIS.

Three possibilities in respect of the method of MIS delivery are initially available. Each approach requires an early evaluation in conjunction with the specification development, to decide which is the most cost effective for MIS implementation. They are:

1. In-house software design and development. This approach has several advantages which are (Morris *et al.*, 1992):
   (a) Written for a known specific application within the company by company personnel, which allows the stepwise development of the MIS within a designated project timescale. Control over the cost of development and easy modification to the MIS if required after user experience.
   (b) Designed to run on the hardware currently used within the firm, and hopefully, on any foreseeable upgrades to the company computer system.
   (c) Written with the full knowledge/active collaboration of company maintenance and production personnel, thereby ensuring acceptance and use of the MIS, by catering for individual and particular requirements in the design of the system.

   The approach also has some disadvantages which are:
   (a) Expense and development time over commercially available MIS. Not necessarily a disadvantage in large organizations with many, possibly unique applications and for which a commercial MIS is unsuitable.
   (b) Re-invention of existing software, or production of software which is simplistic as compared to commercially available MIS packages.
2. The purchase of a commercial package, with if necessary some modification, to allow its possible customization for a particular company application. This approach has the following advantages:

(a) Probably the most time and cost-effective method of obtaining a maintenance information system. For the majority of standard applications, it is possible that some previous user experience can be utilized.
(b) A wide range of commercial software is available, some of which can be tailored to suit a particular company specification/budget limitation.
(c) Comprehensive software surveys are available which detail commercial system features, and which can be useful in deciding MIS specification and selection (CAMM Group, 1988).

The disadvantages of this approach are:

(a) The time and cost involved in the evaluation of competing software. This includes development liaison time, review of software services, license charges, source availability and contractual obligations. All of which are not inconsiderable factors in this approach.
(b) Maintaining control of the cost of development if modifications are required to the standard software.
(c) Ensuring co-operation in development, plus the acceptance and use of the MIS within the company.

3. Contract software design and development, thereby producing a bespoke MIS tailored to the needs of the company. The use of this approach can be beneficial in that most of the advantages quoted above may be maintained while the disadvantages are limited. Employment of independent specialist software consultants will ensure that, via their expertise and knowledge, a state-of-the-art MIS will result. Care will be required however in defining the system specification and project costing, plus ensuring that adequate development and implementation collaboration will take place for system acceptance and use.

From the foregoing it should be obvious that the system specification and its development approach are intertwined, indeed considerable information of use in drawing up the MIS specification can be obtained by investigating the development approach and evaluating currently available commercial software. Another useful technique is the identification of desired system features and their weighted ranking within the system specification. Table 2.1 outlines an example of this type of feature ranking which can also be used for competing system comparison (Beadle, 1989).

2.4 DEVELOPMENT JUSTIFICATION

The justification for purchase, or 'in-house' development of an MIS can involve many factors including the organization of the company and the specification of the MIS concerned. Accordingly this process may be

Table 2.1 Typical example of ranked weighted feature chart

Desired feature	*Importance weight*	*Ideal system*	*Competing Systems A*	*B*
Section 1 MIS system features (weight × rating = score, 10 good, 1 bad)				
1. Asset register	10	100	80	90
2. Plant history	7	70	49	56
3. Failure analysis	6	60	48	30
4. Scheduling/control	6	60	30	42
5. Planned maintenance	5	50	25	30
6. Condition monitoring	5	50	25	25
7. Spare part inventory	4	40	20	20
8. Performance analysis	2	20	10	10
9. Cost analysis	2	20	10	10
Section 2 MIS software features				
1. Modular software	10	100	70	50
2. Language	8			
3. Ease of change	7			
4. Upgrading	5			
5. Servicing	5			
6. Useability	5			
7. Integration	4	40	28	24
8. Experience	4			
9. Standardization	4			
10. Display	4			
Section 3 MIS hardware features				
1. Response time	10			
2. Availability	9	90	90	81
3. Standalone	5			
4. Networked	5			
5. Keyboard	3			
6. Printer	3			
7. Plotter	3			
8. PC type	3			
9. On-line	2	20	10	14
Section 4 MIS environment				
1. Location	7			
2. Protection	7			
3. Authorization	5	50	25	10
4. Ergonomics	5			
5. Training	4	—	—	—
Section 5 MIS Total score/cost		—	—	—

Ideal system rated as 10 on all features (maximum score)

Competing systems subjectively rated by analyst

Table 2.2 Potential cost benefit areas relating to the introduction of MIS

1. Statutory records requirement and information in respect of health and safety inspections.
2. Work order processing, planned preventive maintenance scheduling and condition-based maintenance routing.
3. Plant history, breakdown diagnosis and fault location, reliability analysis, equipment utilization and performance measurement.
4. Spare part inventory control, repair times and logistic delays, material and labour utilization, production losses.
5. Information retention and speed of access, via database generation and search/analysis facilities.
6. Accurate reporting of breakdown situations, maintenance response times, cost and performance data.

viewed as another facet of an MIS project investigation, examination of which is undertaken simultaneously and in parallel with the foregoing areas of organization and specification. Thus the circumstances involved or uncovered by the investigation of any individual and potential MIS application will, in general, dictate what appears in the justification calculation.

However some areas of commonality are possible, and which frequently impinge upon this process, from case to case, when it is undertaken in manufacturing industry. Typical of these are the items shown in Table 2.2, and although the relevance of some of the areas shown may be debatable, their inclusion in any justification calculation is, as previously mentioned, to a greater or lesser extent dependent upon the application (Beadle, 1989).

Industrial legislation for example, and the requirements of statutory bodies such as the Health and Safety Executive (HSE) have to be catered for. Indeed, there is now a greater onus of proof laid on the user of plant and equipment to show that safety legislation has been complied with in the form of regular servicing/inspection etc., when an industrial accident occurs. Such records are easier and more cost-effective to keep within a maintenance information system than on a manual equivalent. However, the amount of cost saving and hence justification for MIS implementation will depend upon the impact of HSE legislation on the firm concerned.

In the same way workplanning information is a common area of interest to many firms in respect of MIS implementation. Accordingly, it is usually a standard feature of most commercial MIS software, however the ability of this feature may vary from package to package and this needs to be checked against the application before purchase. The benefits deriving from computerized workplanning are well documented (Beadle, 1989) and may additionally include savings in some of the areas shown in Table 2.3. The future expansion of simple workplanning into a need for planned preventive maintenance scheduling and/or condition-based maintenance

Table 2.3 Additional cost benefit areas which may relate to the introduction of MIS

1. Although intangible, the strict conduct of safety checks as dictated by MIS control will safeguard against potential accidents and the consequences thereof. These might well include the cost of repairing or replacing physical damage to production machinery, injury to operating personnel and fines imposed by safety legislation.
2. Computerized workplanning may achieve cost savings in addition to those consequent upon a reduced level of administrative personnel. These benefits, which are difficult to quantify, can occur via improved workforce supervision and failure event management thereby leading to a reduction in plant downtime.
3. The manipulation of historical information in respect of plant and equipment will play a major part in improving the efficiency of the maintenance organization and its response to a breakdown. In addition there may well be savings in file space and staff time with such information to hand.
4. A reduction in the spares inventory will obviously lead to a cost saving in respect of the maintenance operation. However additional benefits may also accrue via accurate and timely purchase control, the elimination of waiting time, urgent carriage costs, minimum order requirements, rescheduling and double handling of work.
5. The speed of information access is another obvious benefit of MIS usage together with the availability of current and accurate data. Savings in this area are however difficult to quantify across the complete range of maintenance activity in any typical manufacturing company. The time saved by this feature of MIS may well be considerable and thus provide a significant benefit to the company.
6. The accurate handling of financial data relating to the spectrum of maintenance activities provides more than simple cost control. Such information is useful in the preparation of budgets throughout the maintenance department, and hence assures a measure of performance monitoring against which improvement targets can be set.

routing is an important point to remember under this heading when investigating the capabilities of a MIS package.

The ability to retain and manipulate engineering information is also a fairly standard feature of most commercial maintenance information systems. Again this ability varies from package to package and, as with workplanning, this feature together with its expansion capability should be checked prior to purchase. The benefits of employing a computerized system in this area of maintenance activity are also shown in Table 2.3 and can be extensive. Indeed they are only limited by software ability and requirement, which is why, as will be seen in later chapters, it is so important to ensure that a simple plant history file can be expanded/manipulated to yield information on breakdown diagnosis and fault location, reliability analysis, equipment utilization and performance measurement etc.

Potentially, the ability to access stockholding information via a computerized system can yield some of the largest and most immediate benefits, in terms of reducing the capital cost involved in running a maintenance activity. For this reason it often occupies pride of place in the justification calculation, and rightly so, for savings in this area are by far the least intangible and hence easy to quantify. In addition, Table 2.3 lists some further sectors where savings may well be realized over and above a simple reduction in stock. The expansion capabilities of this feature should also therefore be carefully considered when evaluating a MIS system, for the analysis of repair actions/times, logistic delays, material/labour utilization and production losses may all yield savings via revised operational methods.

Implicit in the foregoing discussion, and thus taken for granted, are the cost savings to be expected as a consequence of replacing a manual maintenance information system with a computerized version. Obvious areas of saving are in the method of information retention and the speed of access provided by a computerized system, which will have database generation plus search/analysis facilities. These capabilities lead to a reduction in the maintenance personnel required to operate the MIS, and faster, more appropriate decision-making based on accurate, refined and timely information.

Furthermore, a key advantage of computerized MIS, which can lead to cost savings, is the ability to handle the financial as well as technical information relating to breakdown situations. Thus a complete analysis of maintenance department response time, cost and performance data for each failure event is possible. With this information to hand, revised procedures can be formulated by management staff to improve maintenance performance and hopefully reduce costs (Antill, 1991).

2.5 IMPLEMENTATION AND DESIGN

The organization, control and management of maintenance in the environment of a modern manufacturing system is probably one of the most difficult, and yet important, of activities which need to be considered, so that both effectiveness and profitability are achieved by the company. Some of the constraints and problems involved in the design and management of practical maintenance systems have already been examined in Chapter 1 and in the foregoing discussion relating to MIS. In this section, a systems approach is used to illustrate within a case study format, how typical maintenance activity, communication and decision flow can be identified and modelled either prior to or in conjunction with, the implementation/ refinement of an MIS.

The company examined has an extensive maintenance system which covers three individual manufacturing plants. It is involved with maintaining a complete range of manufacturing equipment, including ancillary plant and company buildings. The organization outlined in the study is one of eight operating units within a division of the company, and of particular concern to the maintenance department within this unit is a key Flexible Manufacturing System (FMS).

High availability is essential in respect of the FMS, which is in effect a bottleneck within the production plant. This requirement, has in turn prompted the company to investigate the use of 'on-line' Condition Monitoring for the FMS, and also the possibility of utilizing the data so generated, within its existing MIS for management control. This is a good example of how maintenance information systems tend to grow, or develop over a period of time as additional requirements are placed upon them. Rarely is a MIS completely defined, and used 'as designed' throughout its working life without some modification being required, due to either dynamic circumstances or some unforeseen requirement.

In this case, data collection was effected by several computerized acquisition and analysis units fitted to the FMS, and linked to a VAX cluster containing information on production schedules and shift patterns. In turn, the VAX units are connected to an IBM mainframe which maintains all the data relating to equipment and stores inventories. Figure 2.1 outlines the complete maintenance information system, and illustrates the complex interlinking data connections used within the manufacturing unit. The diagram gives some idea of the problems previously outlined with regard to MIS specification, and where in the maintenance organization particular information is required.

This figure also shows a typical mix of existing hardware, which in any company is likely to be used for the implementation of an MIS. Thus by implication, the diagram illustrates typical interfacing problems which in all probability will be encountered during a MIS implementation. A key point to remember is that Fig. 2.1 was drawn at some stage during MIS development and represents the system at that point in time. It is not a design for the complete MIS, but simply one stage in its evolution. Consequently as the drawing did not exist at the specification stage, it could not figure in the MIS design, although had it been available it would have been useful.

A technique which can help at the specification stage, to outline the maintenance organization and identify information required at particular points, is the system modelling and analysis tool SADT. The idea of 'structured analysis' and the concept of decomposition was refined in the early 1970s to produce this technique (Ross, 1977). A subset of the complete SADT methodology, which centres on system functions, is called the activity model and it is this, along with the data model subset, which has been applied to the maintenance system in question.

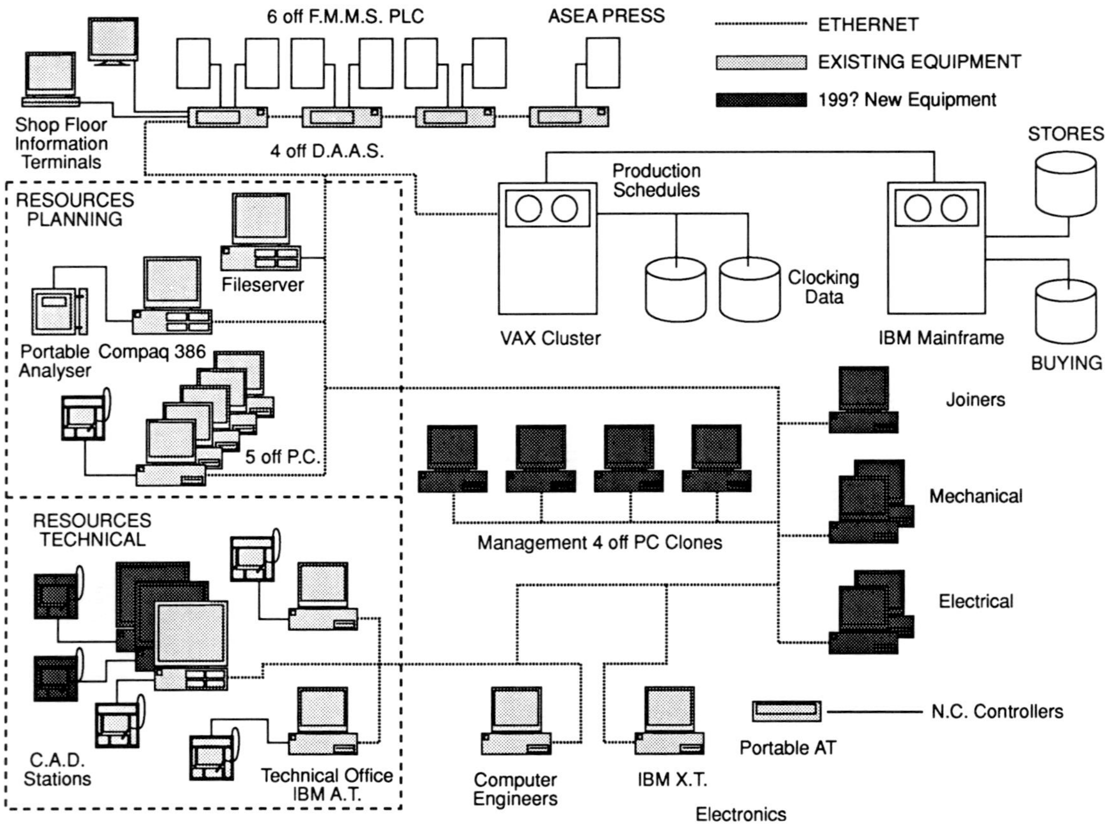

Fig. 2.1 Resources computer system (Morris and Skordis, 1992).

The most fundamental benefit of SADT is that it can be used to 'model' any system, and in this respect, any maintenance system however complicated, can be reflected in a small model which will clearly present the overall objectives. As most of the problems in MIS result from the fact that their design fails to recognize the objective, and the goal they aim to achieve, this technique is a valuable tool to use during the initial concept and specification stage.

In SADT a model consists of both 'natural' and graphical language. The natural language stems from people describing the system, while the graphical language is part of the SADT process. This graphical language manipulates the natural language, such that the resulting system description is far superior to any which would have resulted from the use of natural language alone. The technique, which is incorporated in the SADT methodology can be considered as its most important feature, for it results in a graphical modelling process which involves structured decomposition or the orderly breaking down and addition of extra detail to a complex system which has been reduced to its constituent parts.

Since the method of analysis is top-down, hierarchical and structured, it focuses attention on the important issues and objectives of the system as opposed to the irrelevant ones which are ignored. Decomposition ensures that the design of the system remains 'active' within the initial defined objectives and those which emerge as decomposition proceeds. Thus the system design evolves with known interfaces, security, flexibility and interactively through the definition of inputs, outputs, controls and mechanisms in the form of roles and relationships. Applying the technique retrospectively, as in this case, can also reveal many system deficiencies and consequently suggest modifications for improvement.

As the prime objective of the maintenance system studied was to increase the profitability of the organization it supported, a main requirement of the MIS was that it must function effectively without any unpredictable or intolerable operational characteristics. The current maintenance system had evolved over many years, and epitomized the *ad hoc* integration of maintenance techniques and practices. Hence the need to modernize the structure, achieve optimal performance and integrate it with the introduction of 'on-line' condition monitoring where rapid information flows, analysis and decision-making were key elements in a successful application.

The usual duties of the maintenance system, i.e. maintaining equipment based on breakdown and preventive maintenance philosophies, now had to be aligned with the introduction of predictive maintenance to create a wholly integrated maintenance system. Two problems were rapidly identified as causing operational difficulty and needed to be urgently addressed. These were:

1. The optimal use and exploitation of the existing MIS.
2. The identification and correction of the existing maintenance system shortcomings. These manifested themselves as poor information timing and flow.

A full SADT analysis was performed on the maintenance system in question to resolve these difficulties which involved identifying inputs, outputs, controls and mechanisms in several areas (Skordis, 1990). A typical SADT map resulting from the study is shown in Fig. 2.2, which as an example, outlines in detail the structure of only one area, and one level of the maintenance system within the company. As a consequence of obtaining this type of detailed modelling information, system deficiencies in respect of timing and flow can be identified and were corrected by the company in a revised and upgraded system.

2.6 SYSTEM REDESIGN AND VALIDATION

The SADT analysis conducted in this study revealed an urgent need in respect of improving communication within the maintenance organization, for a great deal of action was based on the type, quantity and flow of information in the structure and the way in which it was stored in the database. Since the database is the key to a successful interactive maintenance system it was vital to identify the properties required of it before offering suggestions for its redesign. In this application, the study defined these properties as follows:

1. The database must support the achievement of the objectives of the maintenance system.
2. The database must interface with, and effectively support if possible, all existing/future systems.
3. The database must itself be supportable and expandable in a distributed processing environment.

Using these properties as a starting point, a logical approach to database redesign was adopted. In order to define the total data requirements of the system, all the various logical paths were grouped using the SADT analysis. A single hierarchy was formed by first using frequency of access and next the priority within the frequency. Thus is was possible to identify and separate information requirements which can only be met by high speed processing, from those which may be serviced by slower and less costly methods.

Data structure diagrams and relational models were subsequently generated, to illustrate the information requirements within the revised system, and these were further analysed with a view to refining the database

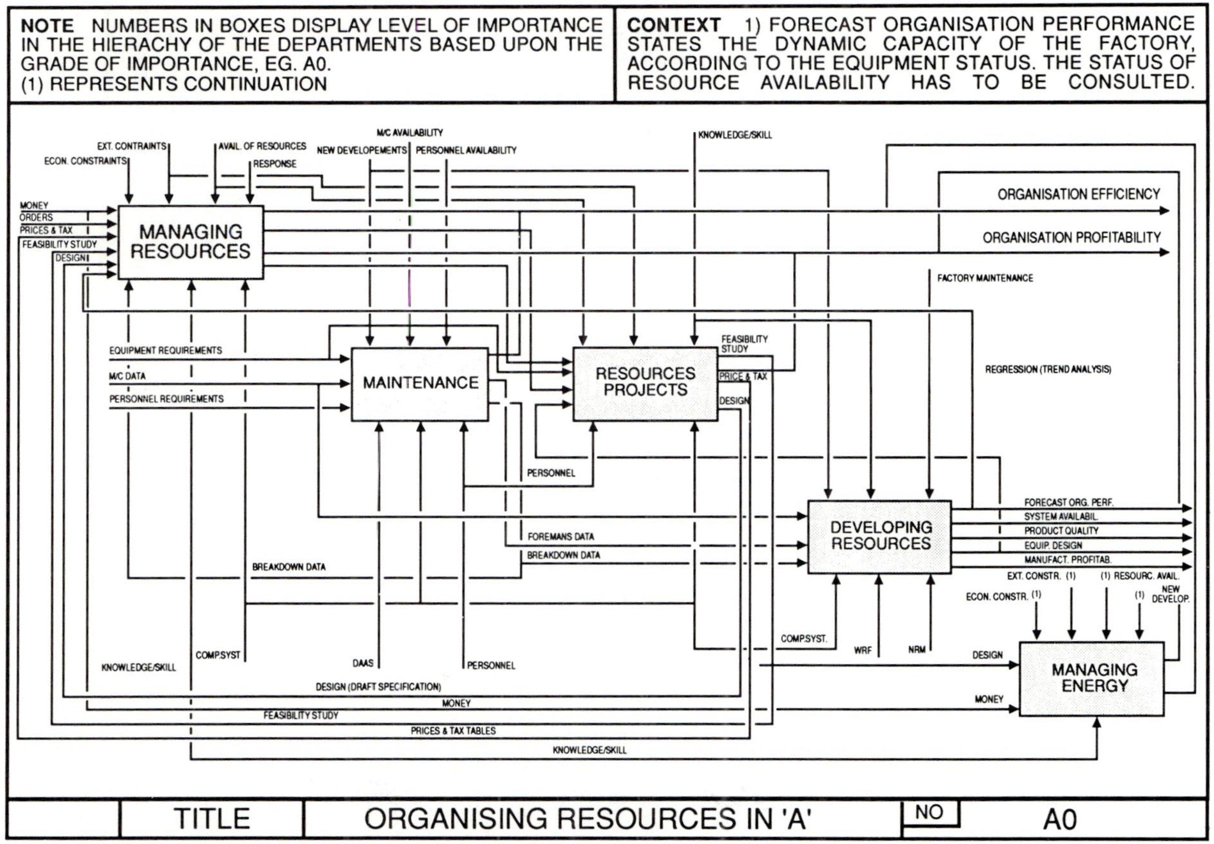

Fig. 2.2 SADT diagram showing resources system in company 'A' (Skordis, 1990).

design. The feasibility of the structure in the redesigned interactive database was then checked using a simplified system employing 'Dbase IV', and use of the Standard Query Language (SQL) to access the information it contained. Subsequent testing, carried out to ensure correct operation under a variety of conditions indicated that the database was operating successfully (Skordis, 1990).

As part of the study, the use of Expert Systems (ES) with this type of database for automatic decision-making was also reviewed, and the conclusion drawn that an SQL driven database could interactively communicate with an expert system although the design would need careful consideration. Successful implementation of this combination requires a clear identification of the overall objectives of the system, and the precise design of the interaction procedures between each software element. Some current MIS systems now offer this feature, where expert system techniques are used to assist managerial decision-making.

The above outline study of a typical maintenance organization shows that a clear definition of objectives is required at the initial design stage. It also reveals that the introduction of extra operational requirements to an existing system, without first considering the overall effect of these changes, can cause grave practical difficulties. In addition, the study illustrates that it is impossible to design an effective maintenance system without first considering the efficient transfer of information through the organization.

Thus, the work done in reviewing this organization shows that the effective operation of a maintenance system is based on the type and amount of data stored in its archive, and that system efficiency is decreased when there is no correct exploitation of this data. Furthermore it outlines how SADT can be successfully used to create the structure of an interactive database, to support a maintenance system, via the modelling of information input and flow. It also indicates that SQL can be used to facilitate access to the information stored, and to provide a link to expert system software, thereby achieving a rapid MIS response in terms of managerial decision-making.

In the full study, special emphasis was placed on the blending of personal computer systems, software and local area networks so that the interactive approach to information access and integrated control was achieved. The real value of interactive MIS was recognized as their ability to aggregate an infrastructure for the immediate and effective access of information. This ability allowed a governing/decision-making body to then control and check the effectiveness of the maintenance organization throughout its operation.

2.7 SYSTEM EXPANSION AND INTEGRATION

Given an increasingly competitive industrial climate, it is highly probable that one of the main characteristics of a successful manufacturing company will be (Tuenschel, 1987):

> The extensive use of highly automated and flexible production equipment.

This implies that the company concerned will have achieved a high level of competence in applying the techniques of Information Technology to manufacturing operations. In such a company economic production is only feasible if:

1. The automated manufacturing system employed is flexible, highly reliable, easily maintainable and thus capable of achieving an excellent level of productive availability.
2. The quality of manufactured output approaches perfection and as a consequence levels of defective product tend towards zero.
3. The company is in essence a 'holistic' manufacturing system, structured to provide maximum and rapid logistic support for production operations via a computerized information network.

In this scenario the maintenance department will have evolved into an efficient support organization, capable of conducting as a matter of routine, any repair or service required on, in all probability, the extremely complicated and high technology production systems then present within the company.

For such a situation to be realized, a radical reorganization may be required in respect of the maintenance engineering carried out by a firm considering the transition to fully automated production. This would involve the areas of repair technique, logistic support, organizational structure, workpractices, staff skills and training. Such a reorganization can be 'driven' by the implementation of Information Technology within a maintenance department and the realization that to achieve successful economic production, the automated factory requires failure free operation. Thus 'failure prevention' becomes the prime requirement of a future maintenance department, superseding both the 'time based', and 'repair on failure' actions which are currently standard practice in many manufacturing support organizations.

Movement in this direction will become inevitable as increased demands are made on the maintenance support function to service a seven day, twenty four hour use of automated manufacturing capacity. Currently this change is reflected in the rapid development and use of Computer-Based Maintenance Systems (CBMS). These, as indicated in Fig. 2.3, utilize the

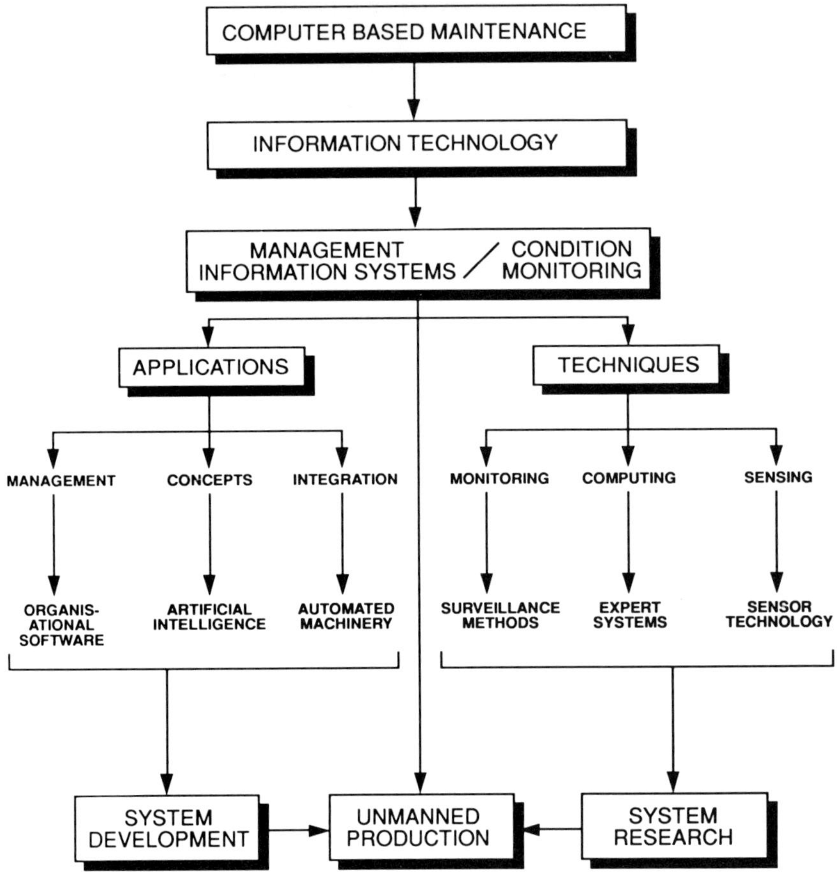

Fig. 2.3 General areas of research and development in condition-based maintenance.

enabling techniques of Information Technology together with the concepts of Artificial Intelligence (AI), to integrate Condition Monitoring (CM) and Maintenance Information Systems (MIS), thereby attempting to provide optimal, expert and predictive management control.

At present there are some difficulties which limit the application of existing CBMS software which include (Davies, 1990):

1. The difficult, unreliable or total lack of communication between different computer systems and/or items of monitoring equipment presently used in automated manufacture.
2. The variation in coding structure used by individual companies to

identify items of production machinery, spare parts, consumables and other parameters essential to the efficient organization of maintenance.

3. The variation in the type and quantity of functional requirements demanded of a CBMS by individual companies, and the consequent complexity this introduces in the design of generic software.
4. The capability to cope with expansion, evolution and integration to ultimately provide a totally computerized maintenance management information and control system, without necessarily incurring obsolescence as technology evolves.

Table 2.4 Some of the benefits of a CBM system within CIM

Computer-based maintenance management system

Condition monitoring and data acquisition system	*Computerized planning and control system*
1. Improved plant availability 2. Improved production control 3. Minimization of scrap components 4. Extended plant life 5. Improved plant safety 6. Rapid identification of potential failures 7. Data for new plant design 8. Lower energy consumption via adaptive control 9. Minimal disruption to production via failure prediction, diagnostics, fault location and computer-aided repair 10. A reduction in maintenance induced failures and the ability to use opportunistic maintenance 11. A reduction in startup time after stoppages for repair 12. Reduced data acquisition and analysis skills 13. An improved accurate compnay wide database 14. Minimal reactive maintenance yielding positive returns on each maintenance action	1. Reduced spare part inventory 2. Reduced purchasing costs 3. Optimum use of maintenance resources 4. A reduction in maintenance personnel 5. A reduction in overtime working 6. Improved maintenance labour force productivity 7. Effective use of contract labour 8. Effective use of clerical staff involved in maintenance 9. Improved documentation, analysis and knowledge of production plant throughout its life 10. Improved operational control and planning of maintenance work/manpower availability 11. Improved communications within other departments within the company 12. Flexibility to add, delete or change information/instructions 13. Hard data available on which to base investment decisions 14. Effective warranty control 15. Holistic approach which covers all plant in the manufacturing system

The envisaged benefits which may accrue from a practical computer-based maintenance system, operated within an automated manufacturing environment, and integrating many of the techniques of predictive maintenance which are examined in subsequent chapters, together with those of maintenance information systems, are shown in Table 2.4.

Analysis of information is the fundamental process which characterizes activity within both these areas, and depending upon the application, this may be done on or off-line, in or out of real time, by manual or computer-based systems. Thus the evolution of CBMS, via the integration of predictive maintenance and maintenance information systems is certain, and dictated by the advent of automated production, which by its very nature, relies totally on the acquisition and analysis of actual condition knowledge.

The successful introduction of totally automated 'just-in-time' production is perhaps one of the greatest challenges currently facing manufacturing industry. In terms of information handling, Computer Integrated Manufacture (CIM) has many facets, some of which may be regarded as more important than others depending upon an individual's viewpoint. However for the 'holistic' concept envisaged by CIM to work effectively, each area of this technology must be integrated correctly and operated efficiently via the timely interchange of computerized information.

The activities which are embodied by maintenance within this type of environment are shown in Fig. 2.4 which outlines the general objectives, functions and requirements of an integrated maintenance system. In practice there are several difficulties to be overcome before such systems can be properly realized and these include (Davies, 1990):

1. The acceptance of the premise that a production facility must be treated as a single complete entity, not only for the purpose of functional use but also for the design of the 'real time', integrated, condition-based maintenance system which is required to support it.
2. Achieving an effective computerized link, between an overall maintenance information system and the many individual condition-based maintenance programmes needed for the monitoring and management of production processes, workparts, tooling and plant.
3. The development of a factory wide database linked to the maintenance information system and subsumed within the overall CIM software, to optimize operational planning, scheduling, plant capacity and maintenance decision-making in a dynamic production environment.
4. The realization that maintenance is no longer an isolated function within a production system, and that highly trained staff will be required to operate and maintain computer integrated production facilities, if a reasonable degree of equipment availability is required.

COMPUTER INTEGRATED MANUFACTURING MAINTENANCE

OBJECTIVES		FUNCTIONS		REQUIREMENTS	
RESOURCE PROCUREMENT	RESOURCE AVAILABILITY	RESOURCE PROCUREMENT	RESOURCE AVAILABILITY	RESOURCE PROCUREMENT	RESOURCE AVAILABILITY
Obtain good system design for availability	Improve strategy and costing in maintenance	Aquisition evaluation	Evaluation of maintenance data	Proposal information	Specification history, tooling investment
Ensure good warranty	Minimise cost bottlenecks, downtime and weak points	Procurement planning	Maintenance project planning	Equipment failure data	Programs, profiles, skill utilisation
Minimise service spares	Balance maintenance capacity	Life cycle evaluation	Identify work requirements and priority	Stockout criteria	Effectiveness processes, manning
Ensure good maintenance provision	Analyse actual condition data for planning maintenance	Purchasing requirements parts, tools, materials	Record, manage and administer data on damage, skills, tools and parts	Stores control personnel and data management	Options, skills, failure information

Fig. 2.4 The role of maintenance in automated manufacture.

The financial rewards which result from the installation of fully automatic production methods, will ensure that problems associated with manufacturing system integration are resolved within the foreseeable future. Hence it may be anticipated that difficulties involved in maintaining such systems will also be addressed during this period, including maintenance information systems, for as previously outlined, maintenance can no longer be ignored within an automatic production environment if good plant availability is to be achieved and a return on capital investment realized.

2.8 SYSTEM EVOLUTION AND MAINTENANCE MANAGEMENT

It is important to note that the majority of maintenance information software, irrespective of its original form, will need to change over time to meet new circumstances and/or extra requirements. In this way such information systems may be thought of as dynamic, evolving into a form which best suits the organization and the manufacturing system it services. Accordingly, when developing an MIS, some consideration should be given to the many initiatives now ongoing in the field of maintenance, and which in the future may become requirements of the system.

Terotechnology for example, as indicated by its definition in Chapter 1, represents an 'over-arching' systems approach to the management of industrial plant and equipment. It is in essence a comprehensive methodology for surveying complex, multi-objective, and long time horizon problems in maintenance engineering. Alternative strategic solutions can be evaluated in this approach through a variety of techniques, to indicate which is the most appropriate according to some given criteria. The concept is 'whole system' and 'whole life' oriented, with relevance to any enterprise whose operations involve the acquisition, use and disposal of physical resources.

A terotechnological system may be regarded as a set of activities and decisions concerned with the selection and provision of permanent physical resources, caring for those resources, monitoring their usage, assembling and analysing information, to ensure an improvement in maintenance system activity and plant design. Thus in practice Terotechnology can be a major contributor towards minimizing life cycle cost, and requires for implementation, joint action by accountants, engineers and managers during all stages in the provision, use and disposal of physical assets.

A terotechnological system is therefore concerned with the acquisition of physical resources which are designed for minimum maintenance, caring for those resources during their operational life and taking the decision to dispose of those resources for minimum life cycle costs. Experience from such activities, fedback to the design, procurement and commissioning functions would then allow the system to 'learn' and hence improve its performance over time.

With the above in mind, Terotechnology can be seen to be essentially a concept which advocates a multidisciplinary contribution to optimizing life cycle costs via the maintenance organization. As such, the number of techniques which can be employed in its effective practice are large, and the benefits claimed to accrue are similarly extensive. In recent years, the advent of better data storage, speed of retrieval, and use in maintenance information or expert systems, has allowed many academic ideas or techniques in maintenance to be implemented. All may be subsumed within Terotechnology, if a wide enough view is taken, and this includes the areas of condition monitoring, logistics engineering, total productive maintenance, asset management and reliability centred maintenance.

The common link between each of these and similar techniques or methodologies in maintenance is the acquisition, manipulation, analysis and interpretation of data. Thus current MIS can be expected to develop into generic software systems which allow implementation of all these techniques, depending upon customer requirements. For example logistics engineering is a technique which relates primarily to the maintenance aspects of weapon systems and attempts to optimize equipment life cycle cost by evaluating the cost-effectiveness of various alternatives as they occur during 'programme' life. It is used particularly in the design and manufacture period to ensure the achievement of specified equipment parameters such as reliability, maintainability and availability.

A wide range of techniques are used in this methodology including reliability, maintainability and failure mode and effects analysis. The evaluation of a given system using this approach determines what logistic support resources are required when the equipment is deployed. This is done by identifying the important servicing criteria or constraints which have an impact on the equipment design. All the decisions relating to service support are evaluated on a life cycle cost trade-off basis using financial and engineering techniques to minimize overall whole-life cost. Included in these calculations are the cost of items such as automatic and built-in test or monitoring equipment for diagnostic use on failure. Thus in civil use, the techniques in logistics engineering are directly applicable to maintenance and can be implemented directly via an appropriate information system (Blanchard, 1986).

Likewise in total productive maintenance, a key element in its implementation is operator training, for when a machine fails the operator is idle and if used to provide 'first-line' maintenance this improves labour utilization. The implication for maintenance information systems is that with condition monitoring techniques and access to 'intelligent' computerized shopfloor equipment, plus the proper training, the operator can become a very effective maintenance technician in addition to their primary role in production (Suzuki, 1987). Similar arguments can be made in respect of the other areas quoted above, subsuming them within Terotechnology and emphasizing the key role of MIS in their effective implementation.

The development of sensors and new sensor systems also has implications for MIS evolution. In many respects this is the key to effective condition monitoring and hence MIS, for without the ability to acquire accurate information, the quantitative monitoring and analysis of production systems would be impossible. Within manufacturing industry, the increasing use of automation has begun to dictate a requirement for 'real time' actual condition knowledge. This information is often related to difficult operating environments, and so new sensors or sensing systems need to be:

1. fast, accurate and self-correcting in operation,
2. intelligent with on board filtering/processing of data,
3. simple in design and rugged in construction,
4. non-intrusive and preferably non-contacting,
5. highly reliable with no increase in system complexity.

Maintenance information software associated with the application of such sensors, and utilizing the concepts/techniques of artificial intelligence/expert systems would be extremely powerful, and more akin to autonomous

Table 2.5 Reasons given to support the development of expert systems within maintenance

1. The repeatability of diagnosis given the same input data for frequent or rare events.
2. The system can contain consensus rather than single source knowledge.
3. The system is portable, always available and not subject to the frailties of human experts such as sickness, holidays and the inability to be simultaneously present at two separate sites.
4. The knowledge base can be updated without affecting the controlling software, although practical systems have yet to be developed which can 'automatically' learn new rules as more data becomes available.
5. The knowledge is automatically safeguarded and not lost with the expert on for example death or change of employer.
6. The proven financial effectiveness of such systems when compared to the cost of using a human expert.

decision rather than information systems. The application and development of this type of computational system to a variety of problems in the area of maintenance is currently ongoing, and condition monitoring software for the intelligent interpretation of vibration signals using these techniques is currently available (Milne, 1989).

The reasons given to support the development of this type of software within maintenance are outlined in Table 2.5, and aside from their use in condition monitoring, applications in maintenance management to provide system fault reporting and repair control, labour reporting and control, equipment spare part inventory control and maintenance data recording and analysis have been reported (Carter, 1985). The combination of information engineering toolkits, fourth-generation languages and relational databases make the provision of user defined and user friendly systems cost effective and affordable to develop.

2.9 CONCLUSION

This chapter has outlined the development, implementation and use of computerized maintenance information systems within a manufacturing environment, together with their enabling and integrating characteristics. The latter are especially important features in respect of an automated production facility where rapid response based on accurate data acquisition and analysis is required. The linkage between such information systems and various other techniques/methodologies in maintenance has also been explored, revealing the central role of MIS within a production support organization.

Research into the use of expert systems in maintenance engineering is shown to be an important part of the development of information technology within a manufacturing context. This is especially true for the successful application of computer integrated manufacture. The developments outlined above, plus those in subsequent chapters, which relate to automation, condition monitoring and maintenance management reinforce this point.

2.10 REFERENCES

Antill, P.J. (1991) Computerised maintenance management systems – the end of the line? *Maintenance*, **6** (4), 15–21.

Beadle, J. (1989) The specification and justification of a computerised maintenance management system. *Maintenance*, **4** (2), 8–15.

Blanchard, B. (1986) *Logistics Engineering and Management*, Prentice-Hall, Englewood Cliffs.

CAMM Group. (1988) *Survey of Maintenance Management Software Packages*, Conference Communication, Farnham.

Carter, J. (1985) Maintenance management – computerised systems come of age. *Computer-Aided Engineering Journal*, **3** (6), 182–185.

Davies, A. (1990) *Management Guide to Condition Monitoring in Manufacture*, Institution of Production Engineers, London.

Gadd, J.R. (1990) Effective use of the computer in managing maintenance. *Maintenance*, **5** (4), 8–10.

Kelly, A. (1991) In search of maintenance organisational excellence. *Maintenance*, **6** (4), 3–14.

Milne, R. (1989) Artificial intelligence for vibration based health monitoring. *Journal of Condition Monitoring*, **2** (3), 213–218.

Morris, E.J., Scott T. and Skordis, M. (1992) Practical considerations for information management. *Proceedings of IEE Seminar on Manufacturing Maintenance*, Coventry.

NEDO Office Report (1985) Advanced Manufacturing Technology – The Impact of New Technology on Engineering Batch Production.

Ross, D.T. (1977) Structural analysis (SA): A language for communication ideas. *IEEE Transactions on Software Engineering*, SE–**3** (1), 16–34.

Skordis, M. (1990) The Use of Systems Analysis Techniques in Modelling a Practical Maintenance System. UWCC (ELSYM) MSc Thesis.

Suzuki, T. (1987) *The Situation of Maintenance in Japan with Particular Reference to the Application of Total Productive Maintenance*. Proceedings of the International Conference on Modern Concepts and Methods in Maintenance, London. Session 1, Paper 3. 1/3/1–1/3/14.

Tuenschel, L. (1987) *The Key Role of Maintenance in Future Manufacturing Strategies*. Proceedings of the International Conference on Modern Concepts and Methods in Maintenance, London. Session 4, Paper 14. 4/14/1–4/14/33.

Zorkoczy, P. (1982) *Information Technology: An Introduction*, Pitman, Bath.

3 Basic diagnostic techniques

3.1 INTRODUCTION

Answering the question, 'What is wrong with it?' is fundamental in maintenance. Conceptually the situation is straightforward; can the cause of the fault be inferred given the available quantitative information? This information can be diverse ranging from the discrete, such as events have or have not occurred to the continuous, such as voltage levels. Here we consider techniques that can be applied to both types of checkout.

Before any diagnostic technique is used the fact that a fault exists has to be established. This requires a clear knowledge of what constitutes the binary divide between sick and healthy operation. For items such as limit switches this is not too difficult but for units whose performance drifts with time this is easier to say than to implement. Typically there is not a clear decision level between these two states. Systems are 'all right, a bit off-colour but we can get by'; the decision level is fuzzy. Manual control of manufacturing systems can be flexible but automated monitoring and control is generally governed by deterministic rules. In this case the binary divide has to be specified. I accept that the use of fuzzy logic (Kaufmann, 1975) can offer a graduated emphasis between these two states but even this approach requires the establishment of the deterministic path between the two states such as the 'S' curve.

In general there are two methods of checkout, one based on a diverse set of tests and one based on a single characteristic such as a dynamic response. The first approach is illustrated in the figure. Assume, for convenience, there are two tests a and b. If we further assume that the tests are not correlated, the test results can be displayed in two dimensions using the orthogonal axes a and b. The x's represent pairs of results that represent healthy systems whilst the o's represent sick systems. In this

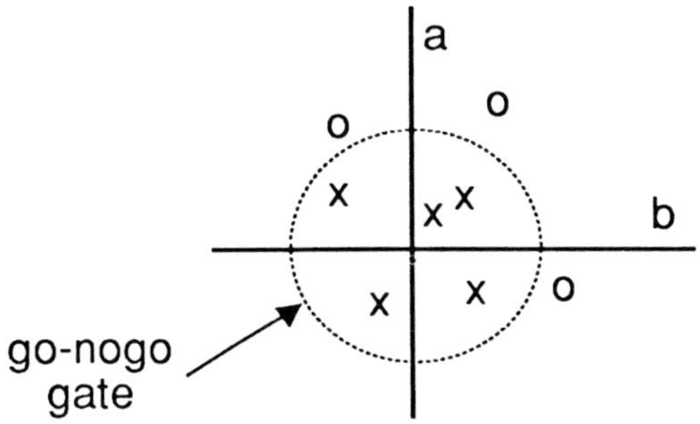

example the circle represents the boundary between the two states. In practice the number of tests will be such that a physical display is impossible, but the methods determining whether or not the 'point' lies within or without the boundary are the same for any number of tests. For example the distance, in two dimensions as in the illustration, of the point (a, b) from the origin is given by $\sqrt{(a^2 + b^2)}$ whilst the 'distance' in five dimensions of a 'point' (a, b, c, d, e) from the 'origin' is $\sqrt{(a^2 + b^2 + c^2 + d^2 + e^2)}$. This distance can be compared to the 'radius' of the boundary thus determining whether the point lies inside the healthy region or not.

If the data to be monitored is a dynamic response then a template is used as shown in the diagram. The template gives the maximum excursion

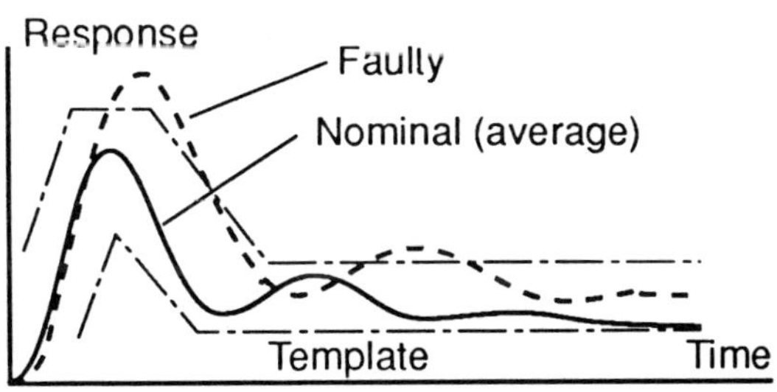

allowable both above and below the nominal or average response of good systems. Then a system that produces a response that cuts the template is deemed to be sick. Adjustment of the template will enable particular constraints to be met. Some systems will have a narrow template because the system being monitored shows little variation whilst in a healthy condition. A relatively wide template will suffice for systems showing more of a variation or where the template is adjusted to allow for natural wear in the system. Care must always be taken in setting the template. If too narrow the likelihood of inappropriate shutdowns increases whilst if too wide the system will run in a sick state.

3.2 QUANTIZED BASIC DIAGNOSTICS

Having established that the system is out of specification the measurements taken are used to diagnose the likely fault. The general method of diagnostic is governed by the nature of the measurements and how the measurements are processed. The general rule for any data processing is the simpler the better. Of course in some situations a simple technique is inadequate and a more complex approach is needed.

3.2.1 Establishing the fault database

Consider the fault indicated in the figure (the dotted line). If the response is sampled at say three points they can be classified with respect to the

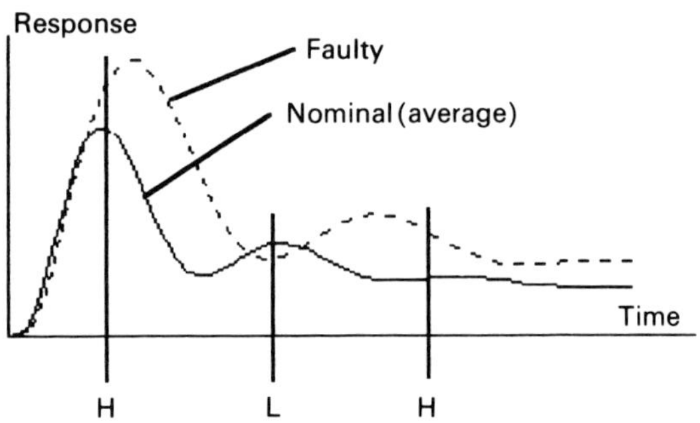

nominal response as high or low. This enables a pattern or fingerprint to be established for each fault (Williams, 1985). The number of combinations of three items taking one of two states is eight as shown:

HHH HHL HLH LHH HLL LHL LLH LLL

One cannot infer that because there are eight distinct patterns, eight faults can be diagnosed. This is putting the 'cart before the horse'. The patterns are generated by the faults. These are then examined to see if the patterns are distinct. If they are not, further samples are taken from the response which will increase the number of items in each pattern but the additional information that the augmented patterns contain may produce a distinct set of patterns. In fact the patterns are not augmented unless the resulting patterns are distinct.

The elements of the patterns or fingerprints considered so far are binary as they take on one of two states, H and L. If the template, which determines whether the system is healthy or not, is used the elements can take on one of three states. The diagram shows a template and two sample points. Depending where the faulty response lies at the sample point, the response

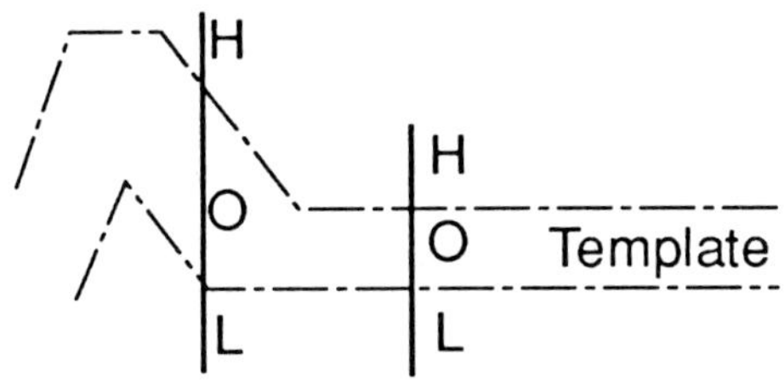

can be high, H; low, L or within, O or, which is more convenient, +1, −1 and 0. Note this classification is now relative to the template and not the nominal response as it was when we didn't use the template. The use of three states can now be formulated by considering the deviations from nominal rather than the actual faulty response. For in practice even in healthy operations the response will show deviations from nominal. It is whether these deviations are large enough to drive the system into faulty operation that needs to be checked. So the deviations, which are always present, need to be checked against the template to generate the corresponding pattern or fingerprint. This pattern is called a signature of the fault. There will be a set of signatures representing the set of likely faults. Collectively these signatures form an array called a fault matrix or fault dictionary.

The fault dictionary is established by simulating a fault and recording the resulting deviation vector. This means that either an actual physical system is available which can be modified to replicate a particular fault or a model representing particular features of the system is adjusted to represent the faults in turn. An alternative to simulation is to record

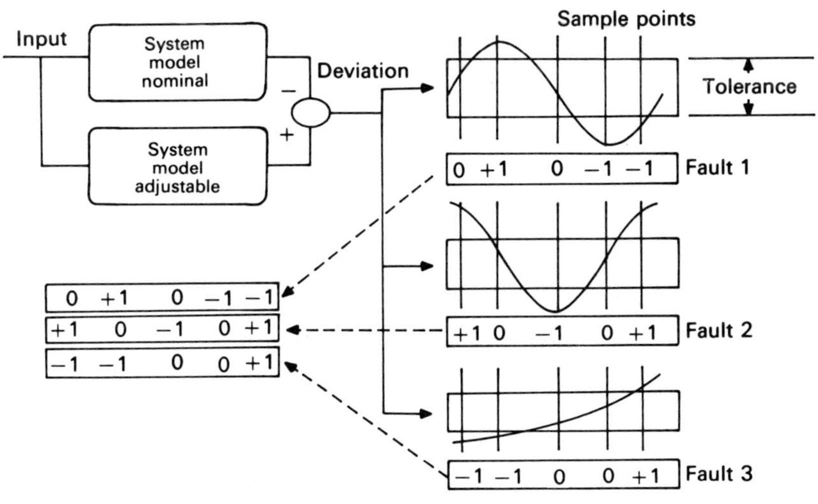

actual failures and resulting measurements in the field. This approach is of little use unless the information is gathered automatically, for usually maintenance records are sparse and it is unlikely that measures of the system at failure are available in normal use. Whatever the method the signatures are collected and stored as a fault dictionary.

The signatures contain the information that represents, to some extent, the shapes of the deviations. It is the shapes of the deviations that can give an indication of the likely efficiency of using them in a diagnostic scheme. If all the shapes are sufficiently different then a good level of diagnosis is obviously possible, whilst similar shapes for different faults will result in confusion and the inability to identify particular faults. Having established the fault dictionary a diagnostic scheme can be established.

3.2.2 Signature matching

The fault dictionary represents the knowledge or memory of the likely faults of the system. When system checkout shows that the system has moved into a failed state the deviation vector is sampled and quantized to produce a signature that is compared to the stored signatures of known faults. The diagram shows the general process. The problem becomes, how are the signatures matched? In the diagram the system under test produced a signature that matched exactly one stored in the fault dictionary. In practice it is unlikely that there will be an exact match for if there are numerous likely faults each signature will contain many elements and the natural uncertainty will produce some mismatching elements. So the problem becomes, how are the signatures matched?

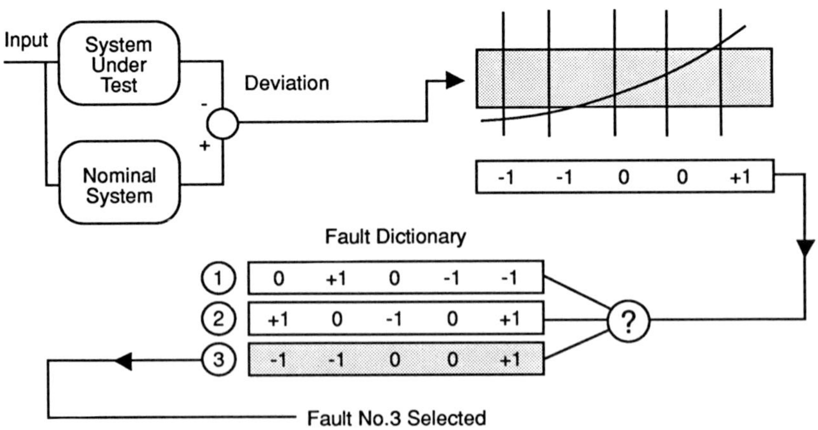

Each fault will have an associated signature. In general we have

$$\text{fault}_j\text{sig} = [a_{j1} \quad a_{j2} \quad a_{j3} \quad . \quad . \quad a_{jm}]$$

These m likely fault signatures can be combined to form the fault dictionary, **A**

$$\mathbf{A} = \begin{bmatrix} a_{11} & a_{12} & a_{13} & . & . & a_{1m} \\ a_{21} & a_{22} & a_{23} & . & . & a_{2m} \\ a_{31} & a_{32} & a_{33} & . & . & a_{3m} \\ . & . & . & . & . & . \\ . & . & . & . & . & . \\ a_{n1} & a_{n2} & a_{n3} & . & . & a_{nm} \end{bmatrix}$$

The system under test will have a corresponding signature which we write as a column vector, **f**

$$\mathbf{f} = \begin{bmatrix} \mathbf{f}_1 \\ \mathbf{f}_2 \\ \mathbf{f}_3 \\ . \\ . \\ \mathbf{f}_m \end{bmatrix}$$

where n represents the number of likely faults and m represents the number of samples taken from the deviation vectors and an unknown-fault signature of m elements. The process of multiplication will produce a vector **b** of the same dimension as **f**, $\mathbf{Af} = \mathbf{b}$. The process is as follows:

$$\mathbf{Af} = \begin{bmatrix} a_{11} & a_{12} & a_{13} & . & . & a_{1m} \\ a_{21} & a_{22} & a_{23} & . & . & a_{2m} \\ a_{31} & a_{32} & a_{33} & . & . & a_{3m} \\ . & . & . & . & . & . \\ . & . & . & . & . & . \\ a_{n1} & a_{n2} & a_{n3} & . & . & a_{nm} \end{bmatrix} \begin{bmatrix} f_1 \\ f_2 \\ f_3 \\ . \\ . \\ f_m \end{bmatrix} = \begin{bmatrix} b_1 \\ b_2 \\ b_3 \\ . \\ . \\ b_n \end{bmatrix} = \mathbf{b}$$

$$b_1 = a_{11}f_1 + a_{12}f_2 + a_{13}f_3 + a_{14}f_4 + \cdots + a_{1m}f_m$$

$$b_2 = a_{21}f_1 + a_{22}f_2 + a_{23}f_3 + a_{24}f_4 + \cdots + a_{2m}f_m$$

$$b_3 = a_{31}f_1 + a_{32}f_2 + a_{33}f_3 + a_{34}f_4 + \cdots + a_{3m}f_m$$

$$b_4 = a_{41}f_1 + a_{42}f_2 + a_{43}f_3 + a_{44}f_4 + \cdots + a_{4m}f_m$$

$$\cdots$$

$$b_n = a_{n1}f_1 + a_{n2}f_2 + a_{n3}f_3 + a_{n4}f_4 + \cdots + a_{nm}f_m$$

The elements of the resulting **b** vector are ranked in descending order.

This re-arrangement gives the order of likelihood of the faults. The first element represents the most likely fault, the next element represents the next most likely fault and so on. The problem of matching vectors using multiplication is the 0 entry. The fact that there is a match between a 0 in both the fault vector and the fault dictionary should increase the count for the particular match, but it doesn't. A better matching scheme is to compare the entries and score a match if a -1 matches with -1, a $+1$ matches with $+1$ and a 0 matches with a 0, and no increase in the count if the entries are not similar, see flow chart. Comparing one element with another is a simple process to automate.

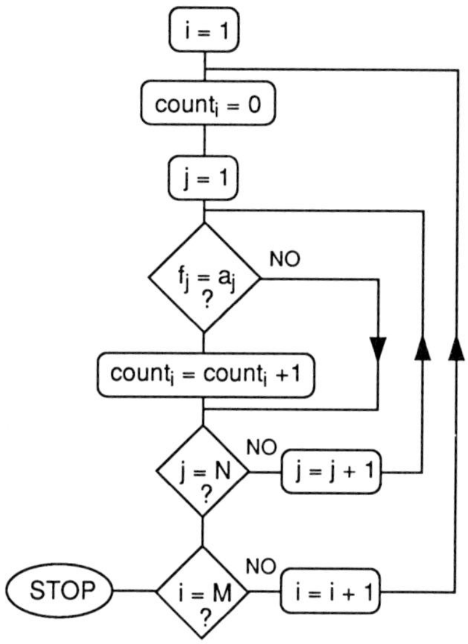

A quantized based diagnostic scheme has advantages and disadvantages. It is a simple technique to apply and quantization can be effective against natural variations in the system response. The effectiveness depends on setting the template correctly. If the noise effect produces variations about the template level then the resulting signature will be subject to variations. The level has to be set so that noise does not alter the signature entries. A disadvantage is that as it stands it does not take into account the level of the fault. As the fault level changes so the signature changes. One proposal is to quantize into many levels. This will mean that there will be a subset of signatures corresponding to a single fault. This is unwieldy except for systems with few likely faults. In addition the very act of quantization means a loss of information or sensitivity. More robust methods based on non-quantized data are available. These are now considered.

3.3 ESTABLISHING NON-QUANTIZED SIGNATURE MATCHING

The transition from quantized based diagnostics to non-quantized based diagnostics is simply accomplished by not quantizing the sampled data. However the data needs to be normalized to allow for the possible wide range of values due to the different drift levels that a fault can reach. As the fault level changes so does the quantized signature. This does not

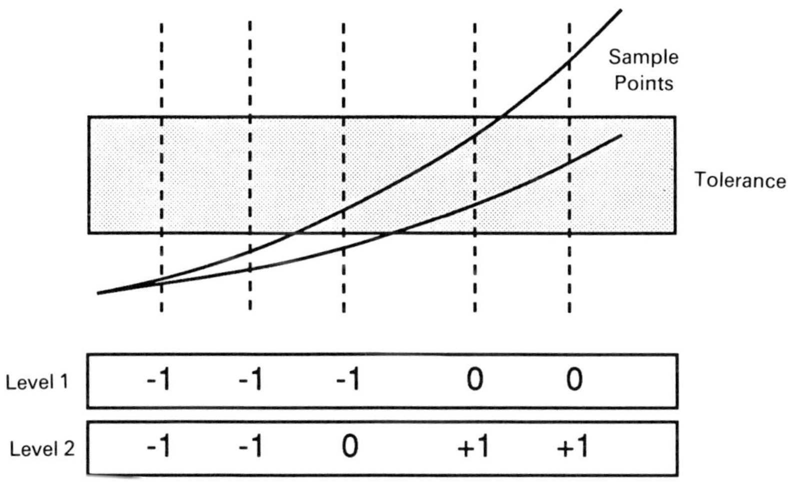

produce a robust diagnostic scheme. One approach is to use multiple entries covering a wide range of fault levels for each likely fault. However, this increases the chance of similar signatures for many faults which are at different levels. The problem can be illustrated if we restrict ourselves to a signature with three elements.

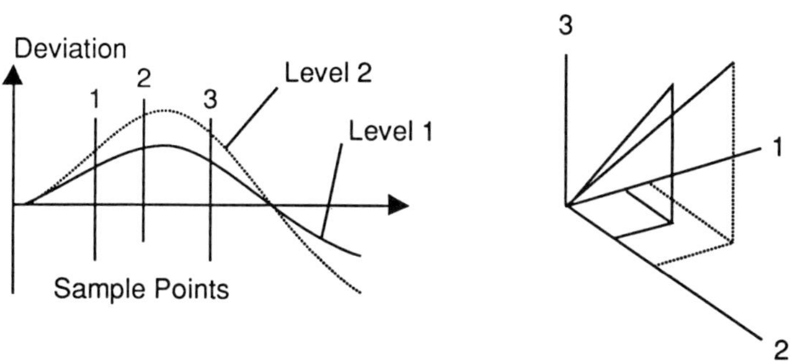

The signature can be thought of as a vector defining a point in three-dimensional space. (In general the signature can be considered as a

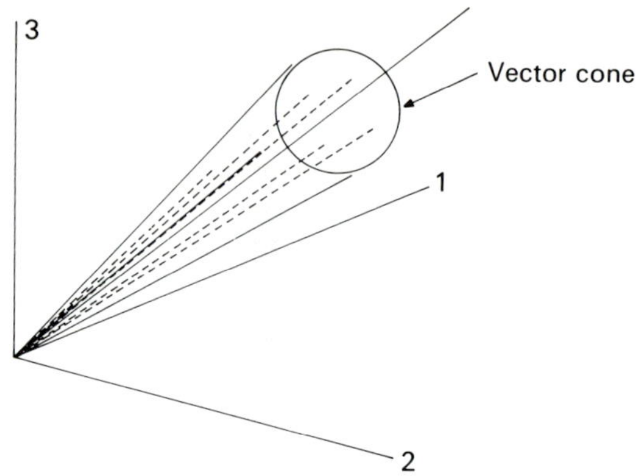

vector in *n*-dimensional space.) As the level of the fault increases the vector changes in magnitude. If we assume a linear relationship between the three elements of the vector the direction does not change. Even if this linear relationship is not present the change in magnitude dominates the change in direction. We can think of a particular fault producing various vectors of varying lengths lying in a 'direction-cone'. So one method of normalization is to represent the vector by directions relative to the three axes rather

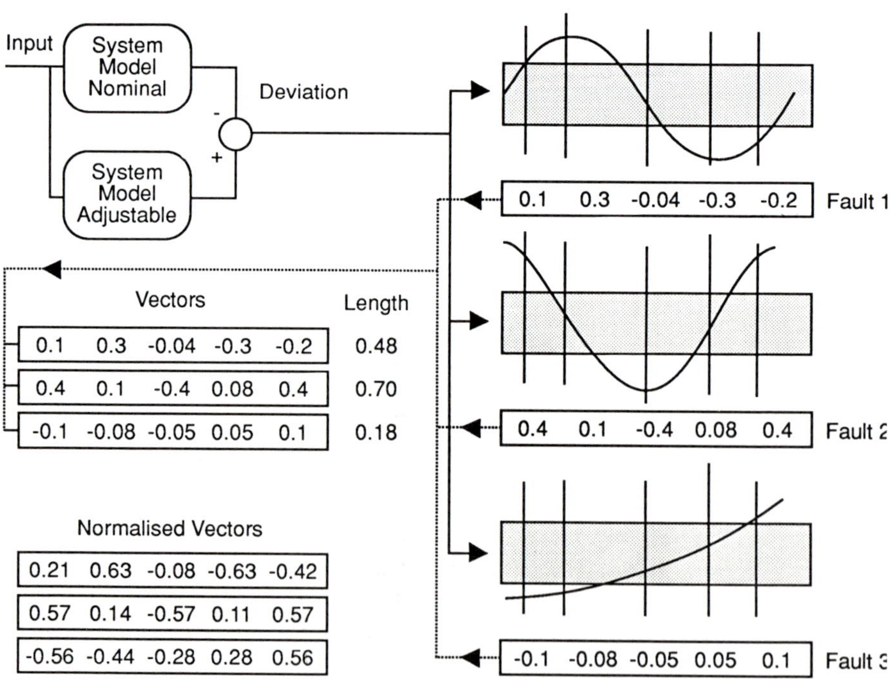

that the elements along the three axes. A convenient method is to divide each vector element by the length of the vector. An element divided by the length is the cosine of the angle that the vector makes with the corresponding axis. In fact the ratio of element value to vector length is called the direction cosine. So the fault dictionary is established in the same manner as in the quantized scheme except that the data values are not quantized but normalized into direction cosines, as shown. Consider a signature **s** where

$$\mathbf{s} = [s_1, s_2, s_3, \cdots\cdots\cdots s_n]^T, \text{ then the length of } \mathbf{s} \text{ is}$$

$$|\mathbf{s}| = \sqrt{(s_1^2 + s_2^2 + s_3^2 + \cdots + s_n^2)} = L \text{ say.}$$

Then the signature **s** is modified to $\mathbf{s}^* = [s_1/L, s_2/L, s_3/L, \cdots\cdots\cdots s_n/L]^T$. The procedure of matching the fault vector with one of the stored signatures is different from the quantized scheme. The method used is to use a 'nearness' measure (Williams, 1985). Nearness can be measured by the distance between the vectors. The distance between two vectors **a** and **b** is given by:

$$d(a, b) = \sqrt{[(a_1 - b_1)^2 + (a_2 - b_2)^2 + (a_3 - b_3)^2 + \cdots + (a_n - b_n)^2]}.$$

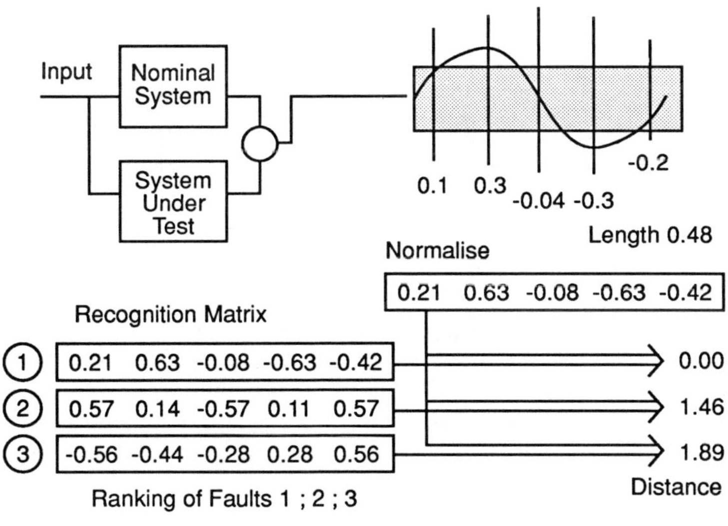

Thus the scheme is to normalize the fault vector and then calculate the distance between this vector and the stored signatures. Finally the distances can be ranked in ascending order which gives the likelihood of the corresponding faults. This scheme is often called the nearest neighbour method.

3.4 USE OF SENSITIVITY FUNCTIONS

An alternative method is based on sensitivity functions and the minimization of an error function (Williams, 1985). Consider:

$$S_j = \sum_1^n \left(y_i - \frac{\partial f_i}{\partial x_j}.\Delta x_j \right)^2 \quad j = 1,2,\ldots m$$

where, y_i = the ith data point of the deviation from nominal of the system under test

$\frac{\partial f_i}{\partial x_j}$ = value of the sensitivity function at the ith point with respect to the jth fault.

Δx_j = the assumed actual change in the jth fault.

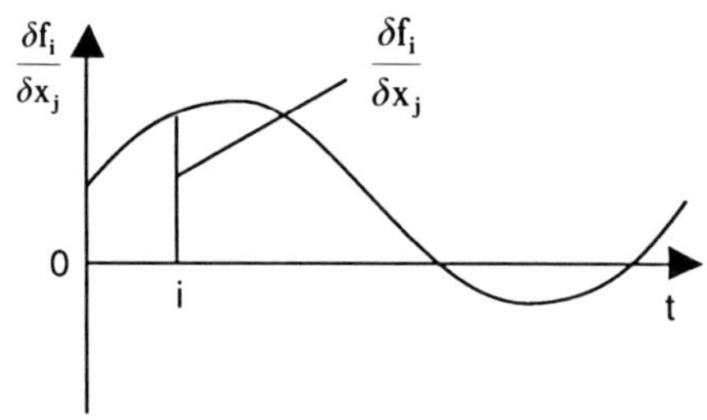

The test domain can be time or frequency and for convenience is indicated by f. The first term in the summation is the actual set of deviations that result from the failed system. The second term is the product of the sensitivity function with respect to a particular parameter and the change in the assumed parameter. In other words, the bracket contains an actual deviation minus the predicted deviation, then squared and summed over the number of sample points. This is repeated substituting a different sensitivity function and the corresponding parameter. Thus we end up with a set of values of S which represent the sum of error squared for each likely fault. The most likely fault will be the one that minimizes S. Thus S is differentiated with respect to Δx_j and set to 0. This gives:

$$S_j = \sum_1^n \left(y_i - \frac{\partial f_i}{\partial x_j} \Delta x_j \right)^2$$

$$\frac{\partial S_j}{\partial \Delta x_j} = 2\sum_1^n \left(y_i - \frac{\partial f_i}{\partial x_j} \Delta x_j \right)\left(-\frac{\partial f_i}{\partial x_j} \right) \text{ for a min. } \frac{\partial S_j}{\partial x_j} = 0$$

$$\sum_1^n y_i \frac{\partial f_i}{\partial x_j} - \Delta x_j \sum_1^n \left(\frac{\partial f_i}{\partial x_j} \right)^2 = 0$$

Thus we have:

$$\Delta x_j = \frac{\sum_1^n y_i \frac{\partial f_i}{\partial x_j}}{\sum_1^n \left(\frac{\partial f_i}{\partial x_j}\right)^2}$$

substituting back into the expression for S_j we get

$$S_j^{min} = \sum_1^n y_i^2 - \frac{\sum_1^n \left(y_i \frac{\partial f_i}{\partial x_j}\right)^2}{\sum_1^n \left(\frac{\partial f_i}{\partial x_j}\right)^2}$$

The first term on the right hand side is independent of the fault considered (j) as it is the sum of squares of the deviations of the system under test. Thus for a minimum of the whole expression the **second** term must be a **maximum**. This gives a method of ranking the likely faults. Calling the second term I_j as

$$I_j = \frac{\sum_1^n \left(y_i \frac{\partial f_i}{\partial x_j}\right)^2}{\sum_1^n \left(\frac{\partial f_i}{\partial x_j}\right)^2}$$

the I_j are ranked in descending order giving the order of likelihood of the faults.

To illustrate the method, the table lists the sensitivity functions for four possible faults of a dynamic system.

Sensitivity functions $\frac{\partial f_i}{\partial x_j}$, $i = 1, 2 \dots 10$; $j = 1, 2 \dots 4$

j i =	1	2	3	4	5	6	7	8	9	10
Fault 1	0.15	0.16	2.22	2.80	3.12	2.55	1.85	0.98	0.55	0.30
Fault 2	3.65	2.88	2.24	1.95	1.45	1.15	0.75	0.64	0.46	0.32
Fault 3	1.24	0.45	−0.55	−1.56	−0.98	−0.05	1.22	1.88	2.12	2.66
Fault 4	−2.42	−1.83	−0.88	1.00	1.68	2.20	1.24	0.45	−0.55	−1.20

The system under test gave the following deviations

SUT deviations, y_i, $1 = 1, 2 \dots 10$

SUT i	0.11	0.20	2.30	3.12	2.86	2.12	1.56	0.76	0.45	0.40

The calculations yield

	Likely faults			
i =	1	2	3	4
$\sum_{1}^{10}\left(\frac{\partial f_i}{\partial x_j}\right)^2$	33.8388	35.9281	26.5263	22.1247
$\sum_{1}^{10} y_i \frac{\partial f_i}{\partial x_j}$	32.2180	20.7899	−0.6618	11.4755
I_j	0.9524	0.5787	−0.0249	0.51867

Ranking the I_i's in descending order gives 1, 2, 4, 3. So the most likely fault is no. 1, then no. 2 and so on.

The method is based on the predicted fault deviation being generated by a sensitivity function. In system design the use of sensitivity functions is a standard method of estimating the effect on system performance due to variations or tolerances in the component values. So often they are available either from an analytic analysis of the system or from a simulation study. The drawback of the system is that the changes in parameter values are small for the applicability of the sensitivity functions. Often the **actual** sensitivity functions have to be established by imposing a fault in a representative system, either an actual system or a mathematical model of the nominal system. Of course, the actual system would have to be a 'typical' example of those manufactured. The test domain is not restricted. Historically the frequency domain was popular in both design and test but developments in instrumentation make the time domain analysis equally possible. Sensitivity studies based on mathematical descriptions of the dynamic system are now readily achievable due to the robust modelling and analysis software packages available.

3.5 FREQUENCY-BASED FAULT CURVES

Finally we look at a method based on the loci generated by varying the values of the system components or parameters being monitored. An early method was based on the frequency response of the system (Martens and Dyck, 1972). The frequency response of a dynamic system can be realized as a complex number which has a real part and an imaginary part (Dorf, 1980).

If the dynamics of a system is represented by a transfer function of the form

$$G(s) = \frac{K.p(s)}{q(s)}$$

it can be shown that the response of the system to a sinusoid input is given by

$$G(j\omega) = K\frac{p(j\omega)}{q(j\omega)}$$

where ω is the frequency and $j = \sqrt{-1}$. $G(j\omega)$ is a complex number of the form

$$\frac{a(\omega) + j.b(\omega)}{c(\omega) + j.d(\omega)}.$$

The $a(\omega)$, $b(\omega)$, $c(\omega)$ and $d(\omega)$ in general are polynomials in ω, the coefficients being dependent on the system's parameters. Suppose we isolate one particular parameter p_i. Then this ratio can be rewritten in the form

$$\frac{A_i(j\omega) + p_iB_i(j\omega)}{C_i(j\omega) + p_iD_i(j\omega)} = T_i(j\omega) \text{ say.}$$

This expression can be viewed as a transformation of p_i to a point in the complex plane $T_i(j\omega)$. It is a bi-linear transformation. As p_i varies in value (a straight line), $T(j\omega)$ traces out an arc of a circle. As each parameter is selected $T_i(j\omega)$, $i = 1, 2 \ldots n$ traces different arcs intersecting at the point that represents the value of $T_i(j\omega)$ when all the parameters are at their nominal values. In a diagnostic mode the point in the T plane will lie on, or near, one of the curves, thus identifying the fault. This is mechanized by selecting various frequencies, ω's, which show good separation of the fault curves. Then the distances of the test point from each curve are calculated. These distances are then arranged in ascending order relating to the likelihood of each fault. The difficulty of using this method is that

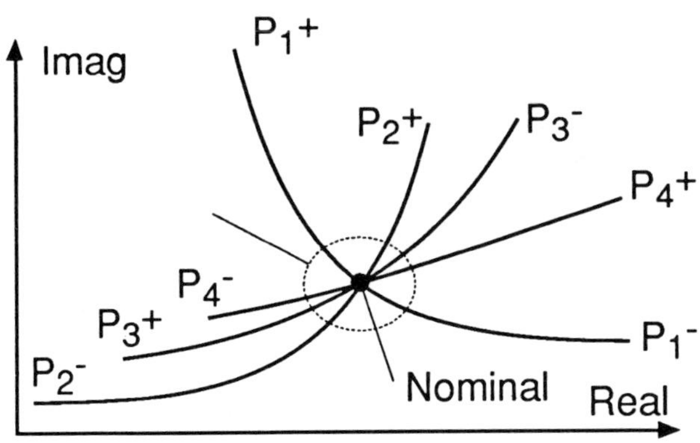

we are restricted to the frequency domain. In CM we are often dealing with operational based information, not a selected test input. However the idea of fault curves is attractive. We now consider a development of the method that is suitable for CM.

3.6 GENERAL FAULT CURVES

Any two measures can be used as the axes for a locus plot for a dynamic system. For example, one could use 'the time to peak' versus 'percentage overshoot' of the system's step response. However, this approach like the Martens and Dyck method suffers from the need of a particular input. Whilst this is possible, in CM it is not desirable. An alternative approach is to use the actual input/output signals which can be processed to produce a dynamic model of the process (Drake, 1991). Then the parameters of the model can be used to monitor the drift in system component values. One convenient set of model parameters are the gain and the poles and zeros of the transfer function description (see Chapter 6). The poles are the roots of the denominator of the transfer function and the zeros are the roots of the numerator of the transfer function. The transfer function can be written as

$$G(s) = \frac{K(s + Z_1)}{(s + P_1)(s + P_2)(s + P_3)}.$$

Here two coordinates at a time can be selected from the three poles (P_1, P_2, P_3) and one zero (Z_1) to construct the planes. So the number of possible planes is six in this example. The values of the poles and zeros will depend upon the values of the physical components of the system. Each pole and zero will depend on diverse combinations, both additive and multiplicative, of these physical components or parameters. When they have nominal values a single coordinate point will be generated in all the possible planes. The effect of these parameter values drifting from nominal on the values of the poles and zeros is to produce curves emanating from these nominal points. The idea of using more than one plane is to increase the fault cover. In general it is unlikely that one plane will give complete cover. This will be illustrated later.

3.6.1 Diagnosing faults

The method of diagnosis is to plot the coordinates of the poles and zeros of the faulty system onto the planes. The fault is identified by observing which fault curve the point falls on. It is unlikely that this point will actually lie on a curve. So we identify the fault by selecting the curve that

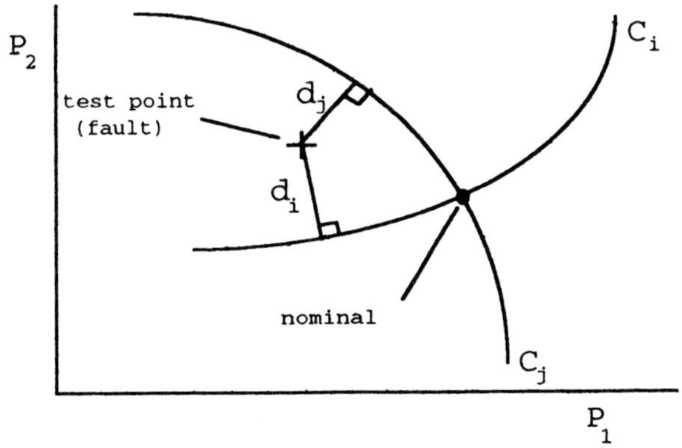

is **nearest** to the point. This is achieved by calculating the distance from the test point to each curve in turn for all the two-dimensional plots used.

Non-dimensional space is used to calculate the distances of the test point from the fault curves, otherwise they would be dependent on the scales used. The following formula is used

$$\text{Normalized distance} = \sqrt{\left\{\left(\frac{x - X}{X}\right)^2 + \left(\frac{y - Y}{Y}\right)^2\right\}}$$

where (x, y) are the coordinates of the test point and (X, Y) is a point on a fault curve. The closest point on the fault curve is the value of X and Y that minimizes this normalized distance. It should be noted that any convenient model is fitted to the fault curve so that Y is 'known' from $Y = f(X)$. An example later will show the variety of useful models that can be used.

3.6.2 Fault probabilities using fault curves

The probability of a particular fault can be heuristically assumed to be inversely proportional to the distance of the test point from the fault curve. If there are *n* likely faults and only one fault occurs at a time then we have:

$$\text{Prob(fault } X_i) = \frac{A}{\text{distance}_i}, \text{ where A is some constant}$$

$$\text{By the laws of probability } \sum_1^n \text{Prob(fault } X_i) = 1 \quad \therefore \quad \sum_1^n \frac{A}{\text{distance}_i} = 1$$

$$\text{So we have, } A = \frac{1}{\sum_1^n \frac{1}{\text{distance}_i}} \quad \text{Hence Prob(fault } X_i) = \frac{1}{\text{distance}_i \sum_1^n \frac{1}{\text{distance}_i}}$$

Thus the Prob(fault X_i), $i = 1, 2 \ldots n$ are ranked in descending order. Problems can occur when some of the fault curves are coincidental. This highlights the problem that as the probabilities of the faults must sum to one the ranking is biased if information on particular likely faults is missing.

3.6.3 Missing probabilities

As we shall see in the example that will follow, sometimes fault curves become coincidental so that we cannot calculate distances. This in turn means that we cannot use the idea of summing the probabilities to unity. This is why we need more than one plane to diagnose all faults. We can estimate the probabilities where this occurs by using the **other** planes. Consider

(a) Single Missing Probability

Let $\text{Prob}_j(i)$ = the probability that the fault is fault i using plot j, where $i = 1, 2 \ldots n$, the no. of likely faults and $j = 1, 2 \ldots m$, the no. of plots used.

Let plot u be unable to diagnose fault k. Then to estimate $\text{Prob}_u(k)$ we use the **other** plots. $\text{Prob}_u(k)$ is estimated by taking the average of the probabilities of the *k*th fault from the other plots.

$$\text{So } \text{Prob}_u(k) = \frac{1}{m-1} \sum_{1}^{m} \text{Prob}_j(k), \quad j \neq u.$$

The other probabilities in the u plot have to be adjusted.

$$\text{Prob}_u(i) = [1 - \text{Prob}_u(k)] \frac{A_u}{\text{dist}_{i,u}}, \quad i \neq k$$

where $\text{dist}_{i,u}$ = distance from fault curve for fault i in plot u and

$$A_u = \frac{1}{\sum\limits_{i \neq k}^{n} (1/\text{dist}_{i,u})}$$

Note that A_u does not include the distance for the undiagnosed fault, k, in the u plot. To compensate for this the probabilities are then multiplied by $[1 - \text{Prob}_u(k)]$ so that the probabilities sum to 1 in the u plot.

The assumption so far is that only one plot has a missing probability. It is not uncommon for more than one plot to have a missing probability. In this situation the adjustments are more complicated. We will now establish a methodology for handling this condition.

(b) Multiple Missing Probabilities

Suppose there are three plots each with one missing probability. For convenience assume that plot 1 cannot diagnose fault 1, plot 2 cannot diagnose fault 2 and plot 3 cannot diagnose fault 3. In plot 1 we cannot estimate for fault 1 from the other two plots for they too are incomplete. Similarly for the other plots. In other words we cannot calculate the adjustment factor $\{1 - \text{Prob}_u(k)\}$.

Let's call the adjustment factors

$$\begin{aligned} a_1 &= 1 - \text{Prob}_1(1) \\ a_2 &= 1 - \text{Prob}_2(2) \\ a_3 &= 1 - \text{Prob}_3(3) \end{aligned}$$

Then averaging in the other two plots we have

$$\begin{aligned} \text{Prob}_1(1) &= \{\text{Prob}_2(1) + \text{Prob}_3(1)\}/2 \\ \text{Prob}_2(2) &= \{\text{Prob}_1(2) + \text{Prob}_3(2)\}/2 \\ \text{Prob}_3(3) &= \{\text{Prob}_1(3) + \text{Prob}_2(3)\}/2. \end{aligned}$$

Substituting into the 'a' expressions gives

$$\begin{aligned} a_1 + \{\text{Prob}_2(1) + \text{Prob}_3(1)\}/2 &= 1 \\ a_2 + \{\text{Prob}_1(2) + \text{Prob}_3(2)\}/2 &= 1 \\ a_3 + \{\text{Prob}_1(3) + \text{Prob}_2(3)\}/2 &= 1 \end{aligned}$$

The probabilities are given by

$$\text{Prob}_2(1) = a_2 \frac{A_2}{\text{dist}_{1,2}} \quad \text{Prob}_3(1) = a_3 \frac{A_3}{\text{dist}_{1,3}} \quad \text{Prob}_1(2) = a_1 \frac{A_1}{\text{dist}_{2,1}}$$

$$\text{Prob}_3(2) = a_3 \frac{A_3}{\text{dist}_{2,3}} \quad \text{Prob}_1(3) = a_1 \frac{A_1}{\text{dist}_{3,1}} \quad \text{Prob}_2(3) = a_2 \frac{A_2}{\text{dist}_{3,2}}$$

These probabilities can now be substituted and the resulting equations expressed in matrix form to give

$$\begin{bmatrix} 1 & \dfrac{A_2}{2\,\text{dist}_{1,2}} & \dfrac{A_3}{2\,\text{dist}_{1,3}} \\ \dfrac{A_1}{2\,\text{dist}_{2,1}} & 1 & \dfrac{A_3}{2\,\text{dist}_{2,3}} \\ \dfrac{A_1}{2\,\text{dist}_{3,1}} & \dfrac{A_2}{2\,\text{dist}_{3,2}} & 1 \end{bmatrix} \begin{bmatrix} a_1 \\ a_2 \\ a_3 \end{bmatrix} = \begin{bmatrix} 1 \\ 1 \\ 1 \end{bmatrix}$$

Thus from this set of equations the a's can be calculated and so the adjustment factors. Subsequently the missing probabilities can be estimated.

3.7 AN EXAMPLE

Consider the seven-component network below:

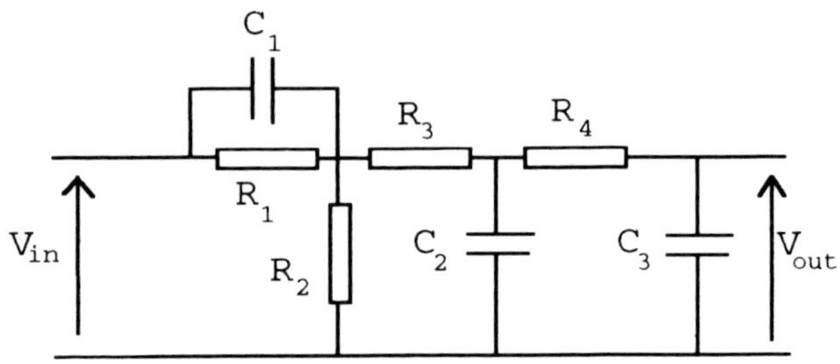

The transfer function is

$$\frac{V_{out}}{V_{in}} = \frac{a_0 + a_1 s}{b_0 + b_1 s + b_2 s^2 + b_3 s^3}$$

where the relationships between the coefficients and the component values are give as

$$a_0 = R_2$$
$$a_1 = R_1 R_2 C_1$$
$$b_0 = R_1 + R_2$$
$$b_1 = R_1 C_1 R_2 + R_2 C_3 R_4 + (C_1 + C_3)(R_2 R_3 + R_3 R_1 + R_2 R_1) + R_1 C_3 R_4$$
$$b_2 = R_1 R_2 R_3 C_1 (C_1 + C_3) + R_1 R_2 R_4 C_3 (C_1 + C_2) + R_3 R_4 C_2 C_3 (R_1 + R_2)$$
$$b_3 = R_1 R_2 R_3 R_4 C_1 C_2 C_3$$

with $R_1 = R_2 = 1\ \text{M}\Omega$, $R_2 = 10\ \text{M}\Omega$, $R_3 = 2\ \text{M}\Omega$, $C_1 = 0.01\ \mu\text{F}$, $C_1 = C_2 = 0.001\ \mu\text{F}$. This illustrates the varied combinations of the component values that make up the model's coefficients and hence the poles and zeros. The model has three poles (P_1, P_2, P_3) and one zero (Z_1) and a gain, K. Thus we have the choice of using 10 possible plots (two from five). It was found that only three were needed. $P_1 v P_2$, $P_2 v Z_1$ and $P_3 v K$. A computer simulation produced the fault curves as shown.

The curves were established by varying each component in turn over the range -60% to $+60\%$ in steps of 10% and recording the values of the poles, zero and gain. From the possible 10 plots the final three were selected visually by the separation of the curves. Fault cover can be itemized by using a table showing the possible discrimination between faults.

The table below shows the discrimination for this example. A '0' entry indicates discrimination and a '1' indicates no discrimination. Element by

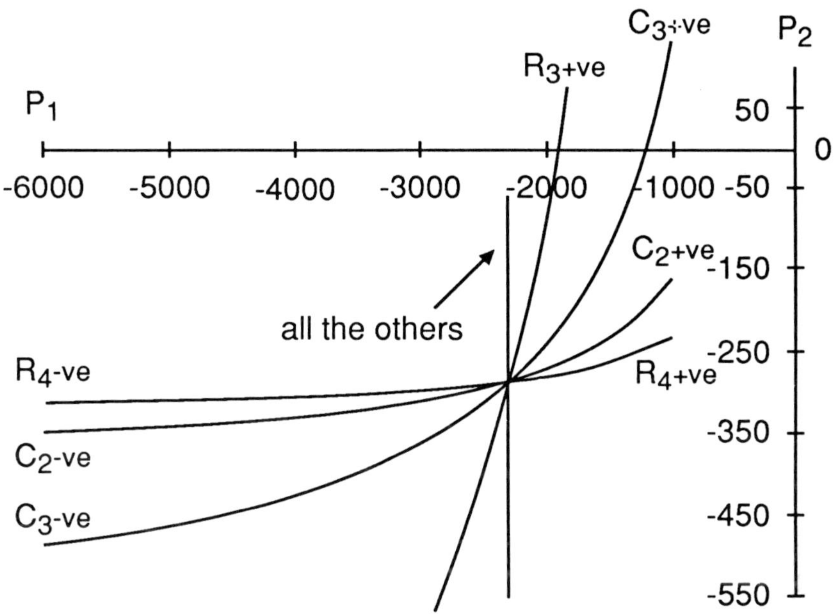

Plot of P_2 v P_1 space

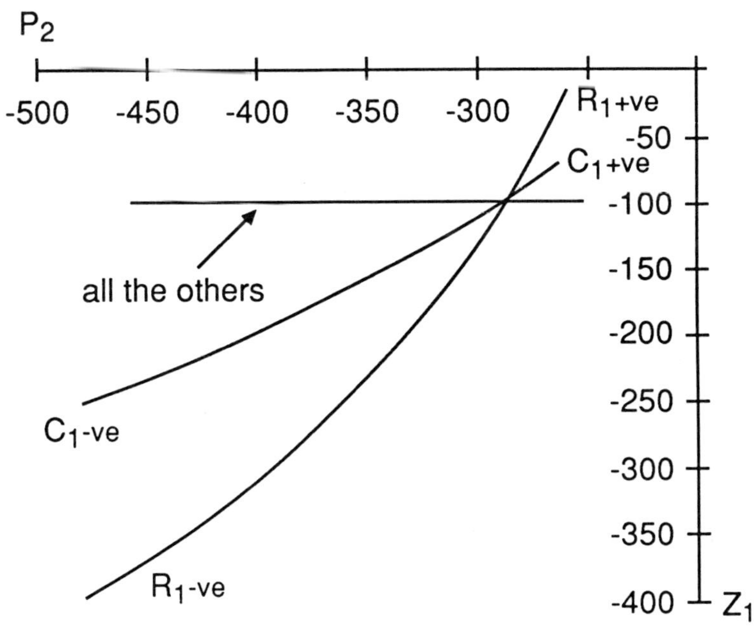

Plot of Z_1 v P_2 space

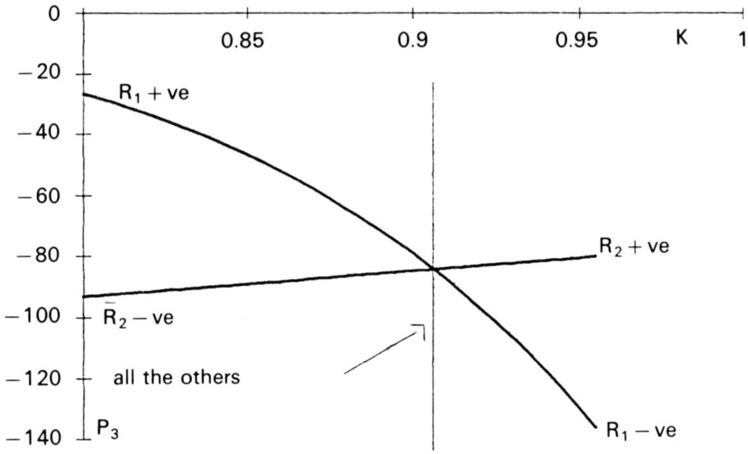

Plot of P_3 v K space

element multiplicaton of these tables produces a table with all entries equal to 0.

	P_2vP_1							Z_1vP_2							P_3vK							Product						
R 1	0	1	1	1	1	1	1	0	1	0	0	0	0	0	0	0	0	0	0	0	0	0	0	0	0	0	0	0
R 2	1	0	1	1	1	1	1	1	1	0	1	1	1	1	1	0	0	0	0	0	0	0	0	0	0	0	0	0
R 3	1	1	0	0	1	0	0	0	1	0	1	0	1	1	0	0	0	1	1	1	1	0	0	0	0	0	0	0
R 4	1	1	0	0	0	0	0	0	1	1	0	0	1	1	0	0	1	0	1	1	1	0	0	0	0	0	0	0
C 1	1	1	1	0	0	0	0	0	1	0	0	0	0	0	0	0	1	1	0	1	1	0	0	0	0	0	0	0
C 2	1	1	0	0	0	0	0	0	1	1	1	0	0	1	0	0	1	1	1	0	1	0	0	0	0	0	0	0
C 3	1	1	0	0	0	0	0	0	1	1	1	0	1	0	0	0	1	1	1	1	0	0	0	0	0	0	0	0
	R	R	R	R	C	C	C	R	R	R	R	C	C	C	R	R	R	R	C	C	C	R	R	R	R	C	C	C
	1	2	3	4	1	2	3	1	2	3	4	1	2	3	1	2	3	4	1	2	3	1	2	3	4	1	2	3

This shows that with these three plots complete cover of the seven possible faults is possible.

Any suitable model can be used for a fault curve because the model is not used to describe the underlying dynamics. The model is simply a convenient representation of the relationship between the two variables of interest. The models used in the above example are:

Space	Fault	Fault curve
P_2vP_1	R_3	$P_1 = \dfrac{1}{-0.000527 - 0.000000316P_2}$
	R_4	$P_1 = \dfrac{1}{-0.00340 - 0.0000103P_2}$

	C_2	$P_1 = \dfrac{1}{-0.00168 - 0.00000431P_2}$
	C_1	$P_1 = \dfrac{1}{-0.000822 - 0.00000134P_2}$
$Z_1 v P_2$	R_1	$P_2 = \dfrac{1}{-0.00392 - 0.00000462Z_1}$
	C_1	$P_2 = -e^{(5.34 - 0.00331Z_1)}$
$K v P_3$	R_1	$K = 0.560(-P_3)^{0.108671}$
	R_2	$K = 1.89 + 0.0117P_3$

3.8 NEURAL NETS

Currently there is much development work being carried out on the use of neural nets (Caudill and Butler, 1992) in diagnostics. Whilst some success has been claimed in their use the development is at an early stage and is not recommended at this time as a robust, easily implemented diagnostic technique for CM. The concept is considered here for completeness and to recommend to the reader to be aware of the method and to anticipate its future use.

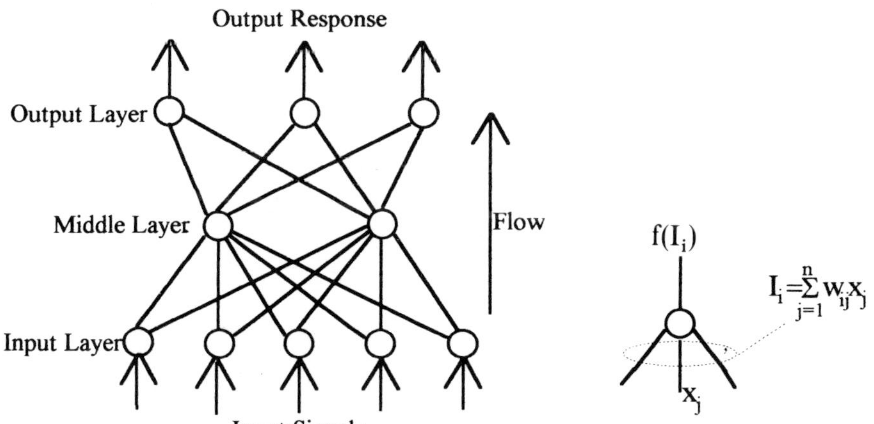

(a) (b)

The fundamental concept of a neural net is simple. It consists of a number of layers of various number of 'nodes'. Every input node is connected to every node in the next layer and so on until the connections are made to the output layer nodes. A typical small-scale example is

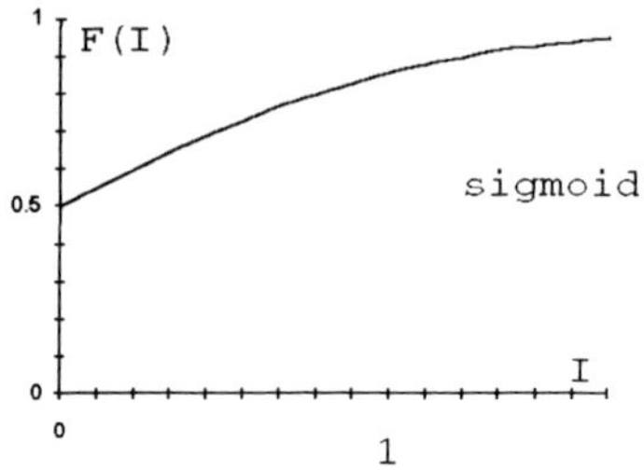

$$F(I) = \frac{1}{1 + e^{-1}}$$

shown. The output from any node is a function of the weighted sum of the inputs to that node. If node i has *n* inputs then a value I_i is computed where $I_i = \sum_{j=1}^{n} w_{ij}x_j$. w_{ij} = weight of *j*th input to the *i*th node and x_j is the *j*th input to the *i*th node. The output of a node is dependent on some function of I_i namely $f(I_i)$. A typical function is the sigmoid as shown. Then if a threshold is set, T say, we can write that the output of the *i*th node is given by: $f(I_i) = f(I_i)$, if $f(I_i) > T$; $= 0$, otherwise.

For a given application the network is trained. This is accomplished by presenting the network with an input signal and a **required** output. The weights are adjusted by minimizing the function which depends on the errors between the actual output and the desired output in an iterative manner. Various techniques are available for this optimization such as hill-climbing (Fryer and Greenman, 1987). In a practical situation thousands of iterations are required and much of current research is focused on speeding-up this process. The field of neural nets is growing rapidly and is likely to yield useful results in the future but there is an element of 'suck-it-and-see' about it.

3.9 CONCLUSIONS

Each application will dictate which method to use. In a CM application the overriding constraint is the need to use existing signals where possible. If this is not a paramount consideration then suitably designed test inputs can enable sophisticated techniques to be utilized. Otherwise the best advice is to use as simple a method as possible to give good fault cover and ease of data processing demands. Popular methods based on conceptual and practical considerations are ones based on the nearest neighbour technique. Whatever method is used it is sensible to update the stored system information at regular intervals, for then the diagnostic scheme will be using information that is pertinent and not based on historical data.

3.10 REFERENCES

Caudill, M. and Butler, C. (1992) *Understanding Neural Networks*, Vol. 1. A Bradford Book, MIT Press.

Dorf, R.C. (1980) *Modern Control Systems*, 2nd ed. Addison-Wesley, Reading, Mass.

Drake, P.R. (1991) Fault location using Relative Pole-Zero-Gain Drift, *J. Syst. Eng.*, **1**, 94–102.

Fryer, M.J. and Greenman, J.V. (1987) *Optimisation Theory*, Edward Arnold.

Kaufmann, A. (1975) *Theory of Fuzzy Subsets*, Vol. I, Academic Press.

Martens, G.O. and Dyck, J.D. (1972) Fault identification in electronic circuits with the aid of bi-linear transformation. *IEEE Trans. Rel.*, **21**, 2.

Williams, J.H. (1985) *Transfer Function Techniques and Fault Location*, Research Studies Press, Wiley & Sons.

4 Vibration monitoring

4.1 INTRODUCTION

The phrase 'running smoothly' is well known and this reflects the fact that for a long time humans have monitored the condition of machines by monitoring the vibrations that they produce. When vibrations are produced acoustic emission is also generated and, if audible, it too is instinctively used by humans to assess machine condition.

Vibration signals carry information about exciting forces and the structural path through which they propagate to vibration transducers. A machine generates vibrations of a specific 'colour' when in a healthy state and the degradation of a component within it may result in a change in the character of the vibration signals. Thus vibration monitoring is a means of monitoring the condition of a machine. Vibration signals can also be used for fault diagnosis since they change in different ways under different fault conditions.

Several advantages are gained by using this condition monitoring method. First, it is well supported by equipment that is readily available and relatively cheap. From an organizational point of view it has the great advantage that once mastered it can be used in a wide range of condition monitoring applications. At the point of application its implementation benefits from the use of non-intrusive transducers. These practical advantages are backed up by the method's well proven effectiveness, the relative ease with which vibration signals can be interpreted, and the reliability of the results.

4.2 VIBRATION MEASUREMENT

Mechanical vibration can be measured in terms of displacement, velocity or acceleration. Low-frequency vibration is best measured using displacement and as the frequency increases so velocity and then acceleration become more suitable. Theoretical considerations show that velocity is the most suitable measurement to use when measuring vibration over a wide range of frequencies. However, the frequency response of velocity transducers is limited, and this prevents their use in high frequency applications. The best all-round solution is the piezo-electric accelerometer because it has a very wide frequency range, even when the signal is converted into velocity or displacement by integration. Another important characteristic of this type of sensor is that it has good linearity throughout its wide frequency range.

However, if very low frequency measurements are required then velocity transducers may be used. Such measurements are usually associated with the monitoring of large and massive systems operating at low speeds. Velocity transducers typically come in the form of seismic detectors that are often heavy, bulky and expensive, so it may be preferable to use accelerometers with integration used to convert the signal to velocity. Proximity sensors may be used to measure shaft eccentricity or misalignment and thereby provide an alternative means of monitoring low-frequency vibrations.

4.3 THE PIEZO-ELECTRIC ACCELEROMETER

Since the piezo-electric accelerometer is the more general purpose and widely used vibration sensor it is the one considered here in more detail. These vibration sensors use a piezo-electric ceramic or pure crystal element sandwiched between a mass and their casing. The sensor may be configured so that the mass exerts either a compression force or a shear force on the piezo-electric element. Their simple design involves no mechanical moving parts so that they are rugged, reliable and not subject to wear. They also have the physical advantages of being small, lightweight and relatively easy to mount. The small charge output from the sensor needs to be amplified and this requires a charge amplifier to be positioned close to the sensor – one may even be built into the sensor's casing. This proximity requirement may preclude this type of sensor from some hazardous applications (Webborn, 1982).

Manufacturers produce a wide range of accelerometers. The principal selection criteria are sensitivity, mass, frequency response, dynamic range and connector position. Special purpose types are available for applications such as: simultaneous measurement in three planes; high temperature environments; very low vibration levels; high level shocks. Miniature types are produced for mounting on delicate structures such as printed circuit boards and for very high frequency measurements. However, a small group of relatively cheap general purpose types will usually satisfy most needs.

The mounting of the accelerometer affects the accuracy of the vibration measurements. Poor mounting reduces the resonant frequency of the sensor system and thereby limits the useful frequency range of the accelerometer. The best mounting is achieved by drilling and tapping a hole on the machine to be monitored and then tightly fixing the accelerometer with a threaded stud. The mounting stiffness can be improved further by applying a thin layer of grease to the mounting surface. When it is not possible to drill a hole, the accelerometer may be glued to the machine although great care must be taken with this approach since it can easily introduce distortion into the signal. Hard glues such as epoxy and cyanoacrylate types should be used to increase the mounting stiffness.

The location of the sensor on the monitored system is also critical to the success of the monitoring exercise. It should be located close to the monitored component to maximize sensitivity and reduce distortion due to the structural path along which the signal propagates. The location problem is synonymous with trying to listen to one person in a crowded hall full of talking people. When a rotating component is being monitored very different results can be obtained according to whether the accelerometer is mounted radially or axially. In any application the user must experiment with different positions to achieve the best compromise between such factors as sensitivity, signal to noise ratio, repeatability, and ease of access and mounting.

When hand-held sensors are to be used, reproducibility as well as accuracy becomes a very important consideration since the sensor should be positioned at the same location every time a measurement is taken. A magnetic mounting can be used provided that there is a good flat surface to take it. O'Sullivan (1991) reports the use of stainless steel discs, permanently attached with glue to the measuring points, to provide a secure magnetic mounting. The discs also ensure that readings are taken at exactly the same location every time. Another solution is to mount a measuring stud permanently on the machine so that when measurements are taken a measuring probe can be 'snapped' on to it. It is important to remember safety when using hand-held sensors so that the operator's hand does not venture into dangerous places.

4.4 OVERALL VIBRATION LEVEL

The simplest way to monitor a vibration signal is to record its root mean square (RMS) average or peak levels. These values can be plotted against time so that any trends can be observed in a similar way to observing product quality trends on statistical process control (SPC) charts. Alarm levels may be set according to the manufacturer's recommendations for the particular machine being monitored, the user's knowledge of the machine or national (e.g. BSI) and international (e.g. ISO) standards. Suitable levels may not be known at first so some initial alarm levels may be set at, say, two times the initial measurement values. These levels can then be modified in the light of experience according to whether they produce alarms too early or too late. Since the measurements are plotted against time, the trend in the signal can be extrapolated, as in Fig. 4.1, to forecast when the alarm levels will be reached. Alarm levels may be set at more than one level. For example, three levels may be set to trigger an increased monitoring rate, a request for maintenance action and shut down respectively. The latter alarm would be particularly important in safety critical situations. It must be remembered that linear trending will not always be appropriate since in many instances the drift towards failure may follow, for example, exponential or 'S' shaped curves.

Often, this very simple method of detecting faults is not very sensitive to the presence of faults until they reach very severe levels. This is particularly true in the case of rolling-element bearings. For such faults the overall vibration levels only exhibit significant increases in the final stages of condition degradation and thus offer late warnings of failure and the need for maintenance.

The vibrations produced by a machine emanate from many sources within that machine and are greatly distorted by its complex structure. This complexity has been described as a 'dynamic cacophony' (Heron,

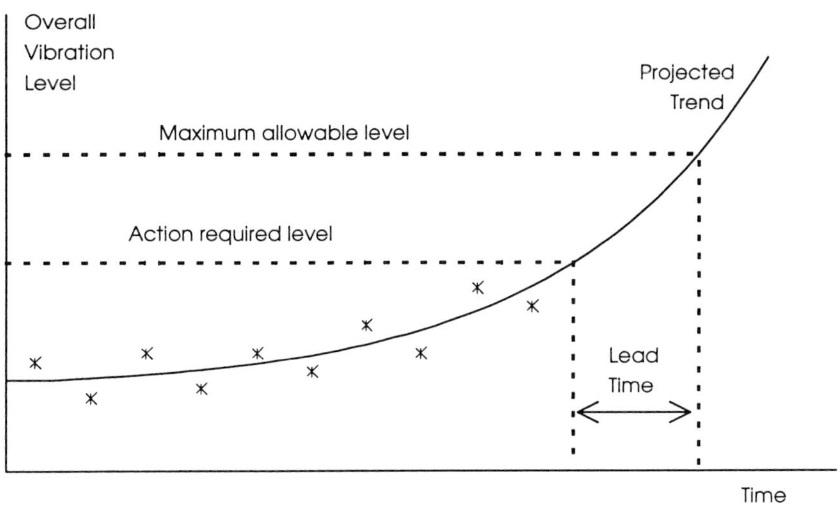

Fig. 4.1 Projected overall vibration level.

1985). In many situations, the components of the overall vibration signal that are indicative of a failing element of a machine can be very small compared to the combined mish-mash of all the components of the vibration signal. This will almost certainly be the case in the early stages of failure, and this is the time at which detection is required for the effective implementation of condition-based maintenance. So what is required are methods for drawing out of the overall vibration signal those components of the signal that detect and diagnose impending failure.

Simply recording RMS or peak measurements against time provides very little information from which to provide a diagnosis of the cause of a predicted failure. This means that, even when this simple method is satisfactory for the fault detection phase of the condition monitoring process, other methods are required for fault diagnosis.

4.5 SPIKY NATURE OF TIME SERIES

In many situations the development of faults will result in an increase in the spiky nature of the vibration time series. This is due to impulses being generated by impacts resulting from damage and is particularly true in the case of rolling-element bearings. Two commonly used methods for monitoring this spiky nature are the Crest Factor and Kurtosis. Both of these signal processing methods have the advantage of producing a single number that is a relative measure of the spikiness of a vibration time series, and this single number can be treated as a 'health-index' that is plotted against time, so that any trend can be detected.

4.5.1 Crest Factor

The Crest Factor is an immediate extension to overall level monitoring and is the ratio of the peak to peak value of a signal to its RMS value.

$$\text{Crest Factor} = \frac{\text{peak to peak value}}{\text{RMS}}$$

Figure 4.2 shows a typical trend for Crest Factor as a rolling-element bearing deteriorates. As expected, the peak to peak value reacts to the development of a fault more quickly than the RMS value. However, as the fault grows in severity, so the RMS value begins to increase, and in the final stages of failure it counteracts the effect of increases in the peak to peak value that typically reach a limit, i.e. the Crest Factor returns towards its initial value. The Crest Factor is much more sensitive than the RMS value alone to changes in the spiky nature of a vibration signal. However, it is much less likely to give sudden false alarms than is the case

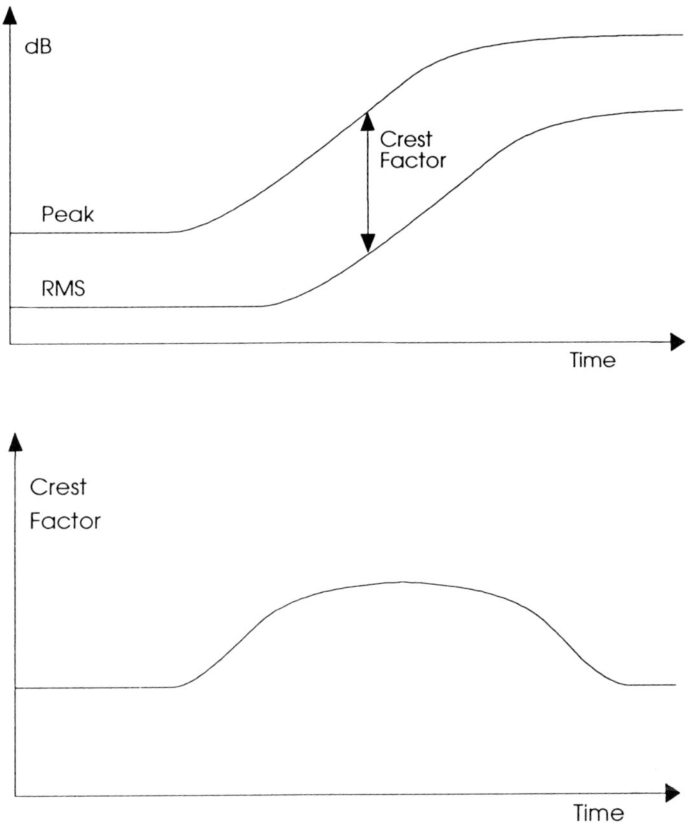

Fig. 4.2 The Crest Factor.

when using the peak to peak value on its own. Also, dividing the peak to peak value by the RMS introduces a form of normalization that takes into account changes in the overall vibration level.

4.5.2 Kurtosis

The Kurtosis (K) of a variable x (in this case the sampled vibration signal) is the name given to its fourth standard moment and is defined as:

$$K = \int_{-\infty}^{\infty} \frac{(x - \bar{x})^4 f(x)}{\sigma^4} dx$$

where: f(x) is the probability density function of x
σ is the standard deviation of x
$\bar{x}$ is the average value of x.

In a practical situation Kurtosis may be estimated as follows for a sample size of n:

$$K = \frac{1}{n\sigma^4} \sum_{i=1}^{n} (x_i - \bar{x})^4$$

As the signal becomes more spiky so the Kurtosis value will increase. The principle underlying the application of Kurtosis to vibration monitoring is that the vibration from an undamaged bearing is not impulsive and has a Gaussian or normal distribution that produces a Kurtosis value of 3. In contrast, the vibration from a damaged bearing is impulsive. This skews the distribution and thereby results in a Kurtosis value greater than 3.

Kurtosis has the important advantage that it is insensitive to changes in machine speed, loading and geometry because these generally affect the whole vibration level and not just one portion of the spectrum. Thorpe and Martin (1991) reported the use of Kurtosis to monitor bearings in the main spindle of a CNC milling machine with much success. In particular, tests showed that when compared with RMS, variance, peak to peak and Crest Factor, Kurtosis is the least affected by temperature and a change of bearing. However, they did find that Kurtosis is not independent of speed when faulty bearings are installed nor is it the most fault-sensitive vibration parameter. The tests verified that for an undamaged bearing Kurtosis is approximately 3 (acceleration distribution is near Gaussian) so that a cusum of (Kurtosis – 3) can be used to monitor the vibration. This cusum gives the advantage of smoothing an otherwise 'noisy' Kurtosis to give a clear and sharp gradient as bearing defects grow. Confirmation that Kurtosis has a fixed value of 3 for a good bearing, and is independent of bearing change, means that Kurtosis monitoring can be readily implemented in a practical situation without having to make adjustments after each machine strip-down and rebuild.

4.6 SPECTRUM OR FREQUENCY DOMAIN ANALYSIS

The overall vibration signal is an amalgamation of a host of components at different frequencies. These frequencies are primarily a function of the dimensions of the machine elements and their operating speeds, and the structural path along which the vibration signals propagate. The signal processing methods considered so far analyse overall vibration levels in the time domain. An overall vibration time series is usually dominated by a few major frequency components such as machine rotation speed. These major components may be important to monitor, but they will mask components with smaller amplitudes that emanate from other machine elements that must be monitored. One way of solving this problem is to decouple the vibration time series into all its frequency components so that the smaller ones can be observed more clearly. The spectra obtained will give much earlier warnings of faults in the machine elements producing the smaller frequency components, since changes at individual frequencies will be observed before the effects become large enough to be detected in the overall vibration time series. Spectrum analysis also enables fault diagnosis since the frequencies at which changes are observed indicate the machine elements that are deteriorating. Even when satisfactory fault detection can be achieved by one of the simpler time domain methods, spectrum analysis is still often required for detailed fault diagnosis.

Alarm levels can be set for the vibration spectra in the same way as they are set for overall vibration levels. As an example, Fig. 4.3 illustrates a reference (good) spectrum with alarm levels (checkout limits) set at 2x and 10x the reference levels. When the current spectrum exceeds the 2x alarm at any frequency then the monitoring rate is increased and the readings at the individual defect frequencies are trend monitored. The 10x check-out limit represents the tolerable limit and if it is reached the machine is shut down. Ideally, planned maintenance should be scheduled before the fault reaches this level. A detailed account of this condition monitoring method is given by O'Sullivan (1991). He suggests that the reference spectrum should actually be a reference mask that is a spectrum broadened at each frequency component in the true reference spectrum. This prevents small changes in speed from indicating vibration increases at the neighbouring frequencies to which the changes move the vibration components. Note that the spectrum presented here is plotted against a logarithmic amplitude scale, i.e. decibels. The use of logarithmic scales is discussed later in this chapter.

Diagnostic charts can be drawn up to relate machine faults to vibration frequencies. These cause and effect relationships can be found by detailed

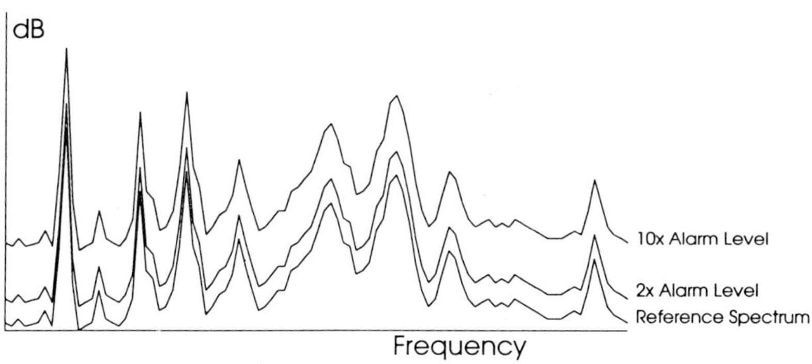

Fig. 4.3 Reference spectrum and alarm levels.

Table 4.1 Vibration diagnostic chart

Fault	*Frequency of dominant vibration*
Rotating members out of balance	1 × rpm
Misalignment and bent shaft	Usually 1 × rpm Often 2 × rpm Sometimes 3 and 4 × rpm
Damaged rolling-element bearing	Impact rates for individual bearing components. Also high frequencies (structural resonances).
Journal bearings loose in housings	Sub-harmonics of shaft rpm, exactly $\frac{1}{2}$ or $\frac{1}{3}$ × rpm.
Oil-film whirl or whip in journal bearings	0.42 to 0.48 × rpm
Hysteresis whirl	Shaft critical speed
Damaged or worn gears	Tooth meshing frequencies (shaft rpm × number of teeth) and harmonics
Mechanical looseness	2 × rpm
Faulty belt-drive	1, 2, 3 and 4 × rpm of belt
Unbalanced reciprocating forces and couples	1 × rpm and/or multiples for higher order unbalance.
Increased turbulence	Blade and vane passing frequencies and harmonics.
Electrically induced vibrations	1 × rpm or (1 or 2) × synchronous frequency.

mathematical analysis (Braun, 1986) or by observing the effects of faults in a live or simulated environment. A typical diagnostic chart for machinery is given in Table 4.1. If any individual frequency component is seen to increase then it can be trend monitored on an individual basis in the same way as the overall level. Waterfall diagrams (3D plots) can be used to trend monitor an entire spectrum. This involves plotting a succession of vibration spectra in a three-dimensional graph where the third dimension is the time at which the spectra are taken.

The Fast Fourier Transform and 'multiple narrow-band filtering' are two methods that are commonly used by commercially available equipment to convert a vibration time series into a frequency spectrum.

4.6.1 Multiple narrow-band filtering

This method can be implemented by using several fixed narrow-band filters which are scanned or a single tuneable filter which is swept across the desired frequency range. The vibration signal measured within each narrow-band is plotted against the central frequency to give a narrow-band frequency spectrum. The filters can be of the constant bandwidth or constant percentage bandwidth types. The former type has a constant absolute bandwidth, e.g. 3 Hz, 5 Hz, etc., whereas the latter has a bandwidth that is a constant percentage of the tuned centre frequency, e.g. 3%, 5%, etc.

The constant percentage bandwidth filter concentrates the analysis effort (increases resolution) at the potentially more crowded low-frequency end of the spectrum. For the same reason, it is compatible with logarithmic frequency scales since on such scales the bandwidths are constant. The value of logarithmic scales is discussed below. However, if harmonic patterns are to be observed, then constant bandwidth filtering and linear frequency scales should be used.

4.6.2 Fast Fourier Transform

The Fast Fourier Transform (FFT) yields a constant bandwidth analysis on a linear scale. It is an algorithm for the rapid calculation of the Discrete Fourier Transform (DFT) of a signal. The DFT is a discrete, finite approximation to the Fourier Transform which decomposes a waveform into its sinusoidal components (frequency and amplitude). A detailed introduction to the Fourier, Discrete Fourier and Fast Fourier Transforms is given by Brigham (1988). The FFT algorithm can be implemented in hardware or software according to the response-time requirements. The FFT is very different to narrow-band filtering in its approach. It is an algorithm which is applied to a sample of the vibration time series. The length of the sample is directly proportional to the resolution of the analysis but indirectly proportional to the full-scale frequency range. A frequency analyser will normally allow the user to define an upper and lower frequency for the analysis and the resolution is then a fixed fraction of this range. This facilitates the very important zoom facility which allows the user to see a more detailed (higher resolution) spectrum in frequency ranges of interest. A typical monitoring procedure would then consist of a broad-band analysis to detect faults followed by zoom analysis to diagnose them in detail.

4.6.3 Narrow-band filtering versus FFT

Direct narrow-band filtering is potentially faster than the FFT. This speed advantage may be required if there are impulsive excitation forces injecting rapid shocks, if the vibration time series is non-stationary or if very fast real-time monitoring is required. Constant percentage bandwidth filters are particularly suited to coping with signal components that exhibit small fluctuations in frequency due to small fluctuations in machine speed. This is because the component frequency is typically proportional to the machine speed and therefore the variation due to machine speed will increase with frequency. Constant percentage bandwidth filters also have the advantage that they yield greater resolution at the low frequency end of the spectrum, i.e. they concentrate the analysis effort on the most crowded area of the spectrum. This characteristic makes them compatible with logarithmic frequency scales which are discussed in the next section.

The FFT on the other hand is better for observing harmonic patterns and providing a very high resolution spectrum. This latter feature is emphasized by the zoom facility that can be vital for detailed diagnostics. In general, FFT is the cheaper alternative, although the gap is currently narrowing. If the FFT is chosen but constant percentage bandwidth spectra are sometimes desired they can be obtained by filtering the FFT spectra, although the speed advantages of direct narrow-band filtering will not be available.

4.6.4 Logarithmic scales and speed compensation

Logarithmic scales are often used for the amplitude and frequency axes of the vibration spectrum. In the case of amplitude, this has the effect of attenuating the smaller components (peaks) in the spectrum. These will not be easy to monitor, on linear scales, if the vibration signal is dominated by one or more very large components that result in the need for a large amplitude scale. It must be remembered that part of the justification for using spectrum analysis is that some of the smaller components of the vibration signal are just as important as the dominant components for fault detection and diagnosis. Another method that has been proposed for accentuating the more important changes in a spectrum, that may be difficult to see due to the relative size of different peaks, is to plot the ratio of the reference (good) and current spectra (Thorpe, 1991). This then gives a plot of the relative change in the vibration across the spectrum.

Logarithmic frequency scales allow a wide frequency spectrum to be plotted on a compact chart whilst still maintaining a high degree of visual resolution, and therefore component separation, in the more crowded low-frequency end of the spectrum. To appreciate this point it must be remembered that a vibration signal is typically made up of many low-frequency components that are related to the running speed of the monitored system (see Table 4.1), but only a few major high-frequency structural resonances, which are excited by impulsive and random forces, so greater resolution is required at the low-frequency end of the spectrum.

Speed compensation can be achieved with the use of logarithmic frequency scales. Table 4.1 shows that many of the vibration components of interest occur at frequencies that are proportional to operating speeds. If linear frequency scales are used then spectra measured at different machine running speeds cannot be compared. However, if logarithmic scales are used, then all the speed-related components of the spectrum will experience the same size shift along the frequency scale when there is a change in speed. This means that spectra measured at different speeds can be compared by aligning them at one particular reference component (peak) in the spectra. The practical application of this method of speed compensation has been reported by O'Sullivan (1991). However, great care must be taken when comparing the spectra produced at different speeds since any changes in the amplitudes may be due to changes in speed rather than machine condition.

4.7 AVERAGING

Averaging may be carried out in the frequency or time domains. The averaging of successive samples of a time domain signal will reduce noise, and will eventually eliminate it if it is statistically stable. Time domain averaging must be synchronized with the forcing frequency (e.g. shaft rotation). This synchronization means that not only the random (white noise) components of the signal will be removed but also those with a fundamental frequency that is not a multiple of the forcing frequency. A spectrum analyser can improve the signal to noise ratio by averaging several time series samples and then converting the average into the frequency domain. However, spectrum (frequency domain) averaging does not remove noise, whether white or at specific frequencies. Instead, it smoothes out the randomness in the amplitudes of the components of the spectra to a mean value. This averaging can be vital as vibration spectrum amplitudes can display substantial randomness.

4.8 SHOCK PULSE METHOD

The shock pulse method (SPM) has been developed for monitoring the condition of rolling-element bearings in particular – one of the most common sources of industrial machine breakdown. SPM can be used to monitor oil film thickness in a bearing and the conditions to support and maintain this film, as well as bearing damage (Allenby, 1988). A defective bearing can be difficult to detect in the normal vibration spectrum until it is very close to the point of failure. The reason for this is that its contribution to the spectrum is relatively small and can be spread across the frequency range. SPM on the other hand measures the mechanical impacts generated by a rolling-element bearing. These impacts generate pressure waves that are transmitted through the bearing housing. These waves will then excite the resonance of a high frequency accelerometer. If the output from the accelerometer is filtered around the resonant frequency the time domain signal obtained will reveal the shock pulses. Filtering at this high-frequency resonance, which is usually at about 32 kHz, effectively filters out the other vibration components that decay rapidly with increasing frequency. This filtering process may be viewed as a means of decoupling the shock waves from the other machine vibrations, or improving the signal-to-noise ratio where noise is considered to be the unrequired 'masking' components of the signal.

The filtered time domain signal can be analysed in several ways. The strength of the filtered pulses, the ratio of the stronger and weaker pulses and the mean rate of occurrence of pulses are examples of simple parameters that can provide the raw data for condition analysis. These parameters can be treated as health-indices and trended in the usual way. The filtered time-domain signal can also be frequency analysed (Shock Pulse Spectral Analysis) to determine the frequency and amplitude of the pulses that can then be related to source and degree of condition degradation respectively.

Off-the-shelf equipment for implementing SPM is available from a company with the same name that has patented the method. The availability of suitable equipment makes SPM a relatively easy to implement method for monitoring bearings. However, it does have its shortcomings. Strictly speaking, the method detects pressure waves rather than vibration and as such cannot be used to monitor the normal or abnormal machine vibrations. Since SPM is detecting pressure waves generated by mechanical impacts in a bearing, measurements must be taken directly from the loaded sector of the bearing housing – there must be no additional interfaces. It has been reported also that SPM, like Kurtosis, is not suitable for slow speed machinery, and it is significantly affected by impact noise from other sources such as conveyors, shakers, vibrating screens, cavitating pumps and reciprocating machines (Smith, 1989; McMahon, 1991).

4.9 ENVELOPING

The impulses generated by a bearing defect excite the bearing housing's natural frequency and other system resonances, i.e. the defect frequency modulates the structural resonances. Enveloping is used in the measurement of this modulating or defect frequency. This is achieved by first band-pass filtering the vibration signal around the structural resonance being modulated. As with SPM, this removes the large low-frequency

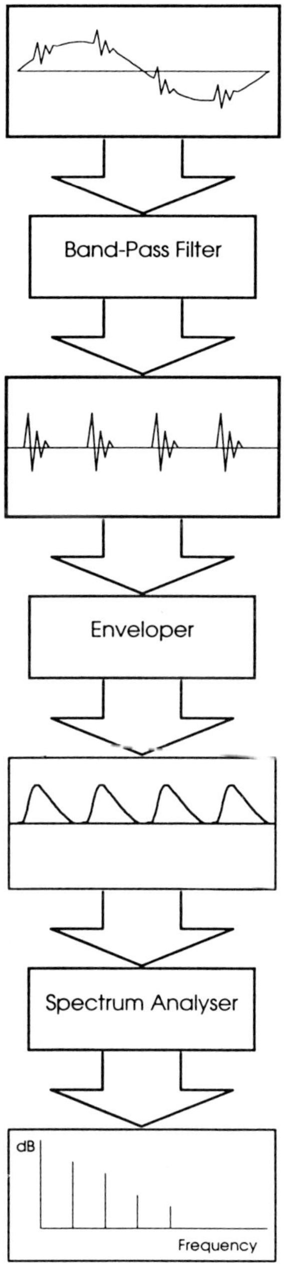

Fig. 4.4 Enveloping.

components of the spectrum. Demodulation or enveloping is then carried out by half or full wave rectification followed by low-pass filtering (an alternative to the low-pass filtering is to use a peak-detector circuit). The demodulated or envelope signal can then be fed into a spectrum analyser so that the defect can be diagnosed in detail. The spectrum obtained will display a harmonic series with a fundamental frequency determined by the characteristic frequency of the fault. In the case of 'rotating faults', the spectrum will also contain sidebands due to the amplitude modulation within the envelope signal (Smith, 1989). The method is summarized in Fig. 4.4.

Detailed descriptions of the method are given by McFadden and Smith (1984) and Ratcliffe (1990). The successful application of the method has been widely reported, for example see McMahon (1991); Thorpe and Martin (1991); Smith (1989). The latter author reported that enveloping, unlike SPM, can be applied successfully to slow speed machines with inherent background noise.

4.10 FURTHER INFORMATION

The manufacturers and suppliers of vibration monitoring equipment often provide excellent booklets and manuals which describe in detail their monitoring methods and equipment. This material is highly recommended for the novice as well as the experienced technician. However, one must remember that the material supplied by one company will only describe the methods and equipment supported by that particular company.

4.11 REFERENCES

Allenby, G. (1988) Condition-monitoring of bearings. *Noise and Vibration Control Worldwide*, July 1988.

Braun, S. (ed) (1986) *Mechanical Signature Analysis: Theory and Applications*, Academic Press, London.

Brigham, E.O. (1988) *The Fast Fourier Transform and its Applications*, Prentice-Hall.

Heron, R. (1985) Condition monitoring as a production tool. *CME*, December 1985, 21–23.

McFadden, P.D. and Smith, J.D. (1984) Vibration monitoring of rolling element bearings by the high frequency resonance technique – a review. *Journal of Tribology International*, February 1984, **17**, 3–10.

McMahon, S.W. (1991) Condition monitoring of bearings using ESP. *Condition Monitoring and Diagnostic Technology*, **2**(1), July 1991.

O'Sullivan, M. (1991) Systematic machine-condition monitoring – a case study from Parenco paper mill in Holland. *Engineering Costs and Production Economics*, **21**, 277–293.

Ratcliffe, G.A. (1990) *Condition Monitoring of Rolling Element Bearings Using the Envelope Technique*. Proceedings of IMechE Seminar on Machine Condition Monitoring, January 1990.

Smith, R. (1989) Effective condition monitoring. *Quality Today*, December 1989.

Thorpe, P.J. and Martin, K.F. (1991) *Milling Machine Bearing Monitoring*, IMechE.

Webborn, T.J.C. (1982) Vibration measurement and incipient failure detection in machines. *Noise and Vibration Control Worldwide*, September 1982, 252–253.

5 Fluid condition and particle monitoring

5.1 INTRODUCTION

The presence of solid particles or other contaminants in lubrication oils and hydraulic fluids can lead to the failure of systems in which the fluids circulate. Degradation of the physical condition (e.g. viscosity) of such fluids can also lead to system failures and reduced life cycles. Furthermore, these fluids carry wear debris from system components which are wearing abnormally or otherwise 'breaking-up'. These are the three major reasons why fluid monitoring is performed.

It can be argued that only monitoring the physical condition of the fluid condition and monitoring for the presence of contamination really predict faults before they occur by detecting the conditions which will cause increased wear and reduced life cycle. If excessive wear debris is present then a fault or abnormal condition has already started to develop, and all that can be done is to take action which will avoid or minimize further damage and to diagnose and correct the fault and its cause.

Spectroscopy is one method available for the monitoring of fluid condition and contaminant since it gives a detailed chemical analysis of the fluid. Modern machinery requires sophisticated modern lubricants which are enhanced by additives. Additives and contaminants can be detected, quantified and monitored by spectroscopy.

The physical condition of a fluid can be measured in terms of such factors as:

- viscosity
- water content
- acidity – Total Acid Number (TAN)
- basicity – Total Base Number (TBN).

If viscosity is lowered then the ability of the oil to separate surfaces is reduced. This results in higher temperatures, oxidation and wear rates. Increases in acidity cause corrosion of such components as piston rings, cylinder bores and anti-friction bearings. Total base number (TBN) measures base additive content. The higher the TBN the greater the ability of the fluid to neutralize acids. This is important, for example, in petrol or diesel engine crankcase lubrication systems were sulphur compounds in the fuel can lead to acidic contamination of the lubricant.

The physical condition of a fluid can be tested in a specialist laboratory at the same time as wear debris tests are carried out. However, kits are commercially available for testing oil samples on-site and these can be supported by 'computer-aided oil analysis'.

Once the condition of the oil is being monitored then a condition-based oil changing policy can be implemented according to additive depletion, contaminant build up and general intolerable changes in physical condition. Such a policy is being adopted increasingly throughout industry and is particularly viable when fluid analysis is also used as a means of system monitoring. The enabling of a condition-based oil changing policy should be viewed as an important financial benefit of the fluid analysis approach to condition monitoring when comparing it with other methods.

As long ago as 1978, the Fluid Power Transmission Group of the IMechE stated that particulate contamination is the major cause of failure of fluid power machinery in industry. A three year survey commissioned by the UK government's Department of Trade and Industry (DTI, 1980) established that there is a clear relationship between fluid contamination level and fluid power system reliability. In modern fluid power systems running clearances are being made as small as possible to achieve maximum efficiency and power, and to reduce the leakage caused by increased system pressures. These reduced running clearances demand increased levels of fluid cleanliness. As a consequence, there is a growing need to monitor for particle contamination in modern fluid power systems to ensure the high levels of cleanliness required. Stringent maximum permissible contamination levels are now being applied to fluid systems to achieve greater reliability. These levels are being achieved by the use of appropriate filtration of the fluids. Therefore, this critical filtering is now a target of condition monitoring by the use of 'post-filter' debris analysis.

The three main causes of particle contamination are:

1. Particle ingress from the atmosphere and dirty workshop environments through seals and reservoirs, and from new supplies of oil which are not as clean as they are required to be.
2. Wear debris generation from within the system. Note, wear debris will always be produced as a matter of course when there is contact between moving parts.

3. Particles retained in the system following construction or maintenance. These particles must be monitored for during the early stages of operation to trigger corrective action such as frequent filter changes until a satisfactory contamination level is reached.

The fluid analysis approach to condition monitoring is restricted to circulatory fluid arrangements such as lubrication or hydraulic power systems. Typical items monitored by this approach include bearings, gears, cams, tappets, piston rings, cylinders and seals. One of the advantages of wear debris monitoring over vibration monitoring is that it can provide a much clearer indication of the source of a fault. The component of a vibration signal caused by a fault can easily be masked by the vast array of large components emanating from perfectly healthy machine components. This is in contrast to wear debris monitoring where, in a fault situation, only the unhealthy machine components contribute excessive amounts of debris. Another advantage is that the path of the debris from source to sampling point is much clearer than that of a vibration signal, and this makes it easier to determine where the 'fault signal' is coming from. It also has the advantage that its measurements can be independent of the speed at which the monitored machine is operating. The disadvantage with wear debris monitoring is of course that it can only monitor those machine components that the fluid comes into contact with. Also, it can be difficult or impossible to achieve a detailed fault diagnosis when using the currently available on-line methods.

The criteria by which a condition monitoring method is selected obviously include cost, ease of use, degree of technical expertise required, accuracy and repeatability. However, there are two other criteria which are particularly important in the selection of a fluid analysis method. The first of these is the speed of turnround of the analysis. The fastest methods will operate in an on-line mode. However, some of the more sophisticated and well established techniques require off-line analysis, possibly off-site at specialized laboratories. In general, there is a trade-off between the analysis turnround and the amount of information provided. On-site analysis will require the provision of greater on-site skills and the purchase of analysis equipment. The use of computerized analysis systems may help in the implementation of sophisticated analysis techniques. The other critical selection criterion relates to what the analysis method can provide information on. For example, some methods can only detect and quantify magnetic particles in a fluid whilst others give a detailed analysis of a wide range of elements which may be present. Also, different methods can detect particles over different ranges of particle size. This means that the size and chemistry of the particles to be detected and analysed must be known before a method can be satisfactorily selected.

Oil analysis may be performed on-line or off-line depending upon the method used. Off-line methods require either particle collection or fluid sampling to be performed prior to the analysis. Particles can be collected from the fluid system by using magnetic plugs or the removal and flushing of filters. Particles which have been separated from the fluid can be examined using optical means such as optical or scanning electron microscopes or image analysis computers. Fluid samples are commonly analysed using spectroscopy or ferrography. In a sense, ferrography can be considered to be a method of off-line particle collection since it separates particles from fluid. The other analysis methods considered here automate the particle counting process either directly, as in the case of light beam blockage sensors, or indirectly, as in the case of methods based on the principle of filter blockage and the consequential changes in fluid pressures. These automated particle counting methods are generally implemented on-line.

It must be remembered that mechanical systems are constantly wearing and producing wear debris. This means that some of the measurements produced by the particle analysis methods will exhibit a continuous rate of increase, rather than a fixed underlying value, when the monitored system is in 'normal' or 'good' condition, and an accelerated (exponential) rate of increase following the onset of a failure, as illustrated in Fig. 5.1. With such measurements it is the rate of change of the measurement which must be monitored since abnormal wear conditions will be evidenced by an increase in this rate of change.

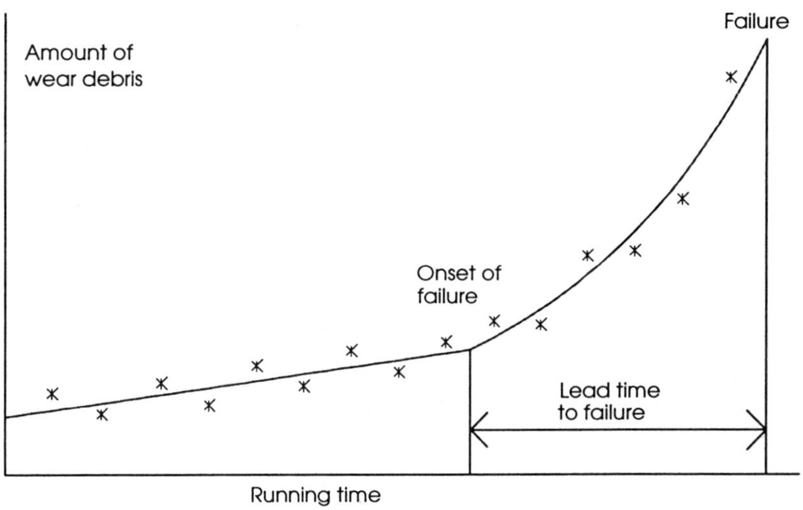

Fig. 5.1 Typical wear debris trend graph.

Further details about commercial suppliers of fluid analysis equipment are given by Davies (1990) and frequent product reviews in such journals as *Industrial Lubrication and Tribology* and the IEE's *Manufacturing Engineer*.

5.2 MAGNETIC PLUGS (MAGNETIC CHIP DETECTORS)

Particle collection for off-line analysis is commonly carried out by the use of removable magnetic plugs which are usually of the quick-release and self-sealing type to prevent loss of fluid. Obviously, this approach to monitoring will detect only magnetic material in a fluid. As such, it is particularly suitable for monitoring ferrous components which produce wear debris in large pieces, e.g. rolling element bearings and gears. It is not suitable for monitoring the general cleanliness of a fluid which will depend upon the presence of non-magnetic material such as silicon. Magnetic plugs are being used more and more in the monitoring of modern jet engines with highly stressed components.

5.3 ON-LINE MAGNETIC PLUG BASED SENSORS

The magnetic plug principle can be automated. In one design, for example, a magnetic plug is placed behind a fluid-tight non-magnetic membrane on which it collects wear debris. The amount collected is automatically measured and when a fixed amount has been collected the magnet is withdrawn for a short time to release the debris. The monitored variable is then the frequency of the withdrawals which will depend upon the amount of wear debris generated. When this frequency reaches an appropriate threshold an alarm is set and a sample of debris is retained for analysis.

Another approach is to place several magnetic plugs close together. If a particle sufficiently large to bridge the gap between two of the magnets is deposited on them an electrical circuit is completed and sets an alarm (Veinot and Fisher, 1989). Such an approach has been developed in the TEDECO QDM (Quantitative Debris Monitor) to count and give an indication of the size of the trapped particles. It is possible that a build-up of fine debris will cause a gap to be bridged and result in a false alarm. To overcome this problem, some implementations apply an electrical pulse to 'burn off' the fine debris without removing the more critical larger debris. A 'burn off' count can be used to monitor the fine debris generation rate.

5.4 PARTICLE COLLECTION FROM FILTERS

The deposits left on filter elements may be used as the basis for analysis. A major disadvantage with this approach is that filter removal and replacement may be difficult and usually requires the machine to be stopped. However, special filter designs are available to overcome these problems. Another disadvantage with this approach is that the filters will not capture wear debris which is smaller than their pore size. However, an advantage is that there is 100% capture efficiency for debris which is larger than the pore size.

5.5 FLUID SAMPLING

When a sample of fluid is taken for off-line analysis the sampling mechanism and procedure is critical to the success of the fluid analysis exercise. The type of particles being monitored for can be microscopically small. This means that the sample holding containers must be scrupulously clean. Reputable oil analysis laboratories will even supply their own clean and capped sample bottles and give advice on taking samples. The sampling point must be chosen carefully. If samples are taken from a tank, then one must ensure that they are not taken from the bottom of the tank where 'gunge' accumulates. Instead, they should be taken from a position which will give samples which are representative of the current state of the system. If sampling is carried out downstream from a filter then large wear debris will not be detected. However, such sampling may be useful for monitoring the effectiveness of the filter – especially in systems demanding very high levels of fluid cleanliness.

Lubricant samples should be taken when the normal operating temperature has been reached and the fluid is in turbulent flow, with the machine still running, to avoid settling of the larger debris. It is important to consider the size of particles anticipated and the ability of the fluid system to transport them to the sampling point. Obviously, samples should not be taken from dead areas of fluid circulation.

The best method of fluid extraction is via a specially fitted tap-off point which should not introduce dead areas in the tap-off line. Tap-off points provide a reliable and repeatable, as well as quick and easy, way to take samples. Reservoir drain plugs should not be used as sampling points as this can lead to artificially high debris readings due to the accumulation of material at the base of reservoirs.

In the absence of tap-off points, syringes or vacuum pumps can be used to extract fluid. Dipstick tubes into the sumps of monitored systems are then often used as sampling points. When syringes are used it is imperative that a new clean syringe is used for each sample to avoid cross contamination by microscopic particles. The same applies to the tubing used with vacuum pumps.

5.6 OPTICAL MICROSCOPE

Particles which have been separated from fluid can be usefully examined using an optical microscope. The first thing this analysis enables is the counting and sizing of the particles which in turn should give an indication of the severity of any wear conditions. Diagnostics can be obtained by classifying the shape of the wear debris viewed under the microscope. Wear Particle Atlases (National Coal Board, 1984; Anderson, 1986) have been produced to relate particle morphology to cause and possible source. Users may build up atlases for their own particular machines by keeping pictures of debris produced under known (diagnosed) fault conditions.

Where particles have not already been separated from the fluid, vacuum filtration can be used and the particles deposited on a membrane with a pore size appropriate to the particular application. The particles can then be sized and counted. Due to the time taken to count all the particles on the membrane, statistical counting techniques may be used.

5.7 SCANNING ELECTRON MICROSCOPE (SEM)

One of the big advantages of SEMs is that they are capable of detecting particles down to 0.1 microns which makes them suitable for contamination control in systems which require very high levels of cleanliness. They can also deal with particles in excess of 1000 microns. The latest SEMs have image analysing programs which enable particle counting to be performed. However, SEMs are fairly specialist pieces of equipment and have high costs associated with them.

5.8 IMAGE ANALYSING COMPUTER (IAC)

The IAC is effectively the computerized version of the optical microscope and involves the image from a video camera being analysed by a computer. The computer can size and count the particles and recreate the image on a viewing monitor. The computer facility enables a more detailed analysis of the particle images to be carried out and can rapidly apply alternative sizing techniques.

Problems may arise with this technique when particles overlap – sophisticated editing techniques may be required to avoid overlapping particles being classified as one larger particle. Focusing can also present problems since various sizes of particles will be present. IACs are a relatively new technology and significant advances are to be expected in this area over the next few years, particularly in on-line applications. Example IAC applications have been reported by Albidewi *et al.* (1991) and Uedelhoven *et al.* (1991).

5.9 SPECTROSCOPY

This well established and widely used method of fluid analysis operates in an off-line mode with fluid samples being analysed either on-site or at remote laboratories. There are several types of spectrometer falling under this general heading.

Spectroscopy has the advantage that it can provide data on the chemical composition of debris, i.e. it can identify metals such as iron, silver, aluminium, chromium, copper, tin, lead, titanium, cadmium, molybdenum, vanadium and nickel. It can also detect lubricant additives such as calcium, barium, zinc, phosphorus, magnesium, boron and molybdenum and contaminants such as silicon, sodium and potassium. Spectroscopy is able to quantify as well as identify the presence of all these elements so that concentration levels can be trended.

The big disadvantage with spectroscopy is that it only detects small particles up to about 5 to 8 microns in size. This is due to a number of factors such as an inability to vaporize larger particles. This inability to detect larger particles can mean that some severe mechanical wear can go undetected since such wear might produce relatively large metallic particles in excess of 10 to 20 microns in size. In a sense there is no lower limit on the size of particles that spectroscopy can detect since it can identify elements in solution.

Spectroscopy is used extensively throughout industry. British Rail have reported great success in its application to monitoring the condition of the engines of High Speed Trains (HSTs). The method is used to monitor the condition of the lubrication oil as well as the engine components which are washed by the oil, and has thus produced significant financial savings by extending the period between oil changes as well as engine overhauls. They have speeded up the process of this off-line analysis by investing in automatically controlled laboratory facilities to handle the tests, and computer and communications networks to handle the rapid analysis and flow of information. The United States Air Force has also implemented an extensive Spectrometric Oil Analysis Program (SOAP) to increase the reliability, safety and flight readiness of its aircraft. Three specific spectroscopy techniques are described below.

5.9.1 Atomic absorption spectrophotometer

In the atomic absorption spectrophotometer a sample of fluid is diluted and vaporized in an acetylene flame. Light with a wavelength corresponding to (absorbed by) the element to be detected is passed through the flame and absorbed if the element is present in the sample. Comparison with a known reference standard provides a direct measure of the amount of each element in the sample. The instrument is accurate, inexpensive, compact and relatively simple to operate.

5.9.2 Atomic emission spectrometer

The emission spectrometer measures the characteristic wavelengths of light which are emitted when the elements in a fluid are excited by a high energy source such as an electrical discharge. Elements are then identified by comparing the measured spectra with reference spectra produced previously by known elements. A direct reading emission spectrometer measures the wavelengths by using photo-multipliers linked to spectral lines and consequently several elements can be detected simultaneously. The equipment is large, expensive and requires precise calibration to produce good repeatability. However, it has the benefit of being able to detect bigger particles than the spectrophotometer. The use of emission spectroscopy for the analysis of wear debris in grease has been reported (Jones, 1989).

5.9.3 X-ray fluorescence

In this method the fluid is irradiated by an X-ray source. This causes energy level changes in the atoms which when reversed emit secondary X-rays. The energy in these X-rays indicates the elements in the fluid sample and the intensity of the X-rays indicates the concentration of these elements. This is a sophisticated method which requires a fair degree of skill in the interpretation of the emitted X-rays. The equipment is very expensive but is reasonably portable.

5.10 FERROGRAPHY

In this method a sample of fluid is brought into contact with a magnetic field in some way which causes the separation of the ferromagnetic and paramagnetic particles from the fluid. The size range detected is from about 1 to 100 microns. Non-magnetic particles can sometimes be captured provided they have had contact with the ferrous parts of the system (Desjardins and William, 1991). However, the big drawback with this method is that in general it is not very efficient at detecting non-magnetic particles.

5.10.1 Direct reading ferrography

Direct reading ferrography precipitates the debris under the influence of an inclined high intensity magnetic field inside a fine bore glass tube. The debris is precipitated at two selected points down the tube. The first of these is close to the entry and monitors the smaller debris. The larger material, typically greater than 5 microns, is detected at the second point. The optical density readouts obtained at these two points, from light sensors, are trend plotted and used to detect samples requiring a more detailed analysis.

5.10.2 Analytical ferrography

In analytical ferrography the fluid is caused to flow over a specially prepared substrate across a magnetic field that separates the ferromagnetic and the paramagnetic particles from the fluid. The larger particles are deposited first with progressively smaller debris being deposited as the fluid flows across the magnetic field. The morphology of the particles separated from the fluid can be determined by the use of optical microscopes, scanning electron microscopes or image analysing computers as described above. The separation patterns may also provide useful information since they are related to the particle sizes and their distribution.

5.10.3 Rotary particle depositor and particle quantifier

This off-line system of wear debris analysis has been developed at the University of Wales, Swansea Tribology Centre (Price and Roylance, 1984) and is effectively an enhanced method of ferrography. The fluid sample is deposited in the centre of a slide using a syringe operated pipette. The slide is held onto a permanent magnet by a rubber suction seal and rotated by means of a variable drive. This causes the fluid to flow radially outwards so that the debris is deposited on the slide as a set of concentric rings. This process provides rapid deposition of wear particles which can then be analysed by the use of an optical microscope, electron microscope or image analysing computer. For high sample throughput and an indication of particle quantity, the rotary particle depositor (RPD) is used in conjunction with the particle quantifier (PQ). This is a magnetic debris monitor which assesses the quantity of particles in a sample in terms of their magnetic density and thus produces a value known as the 'PQ' index. The particle quantifier is intended to provide rapid measurement of particle concentration, and as the index is directly proportional to the magnetic mass of the particles, it does not matter if the sample is held in liquid suspension or is deposited on a slide. The complete particle quantifier product incorporates a computer interface, a keypad, and a printer.

5.11 ON-LINE PARTICLE COUNTING USING INDUCTANCE SENSORS

This method employs an inductance sensor consisting of a cylindrical coil of wire wound around a non-metallic pipe through which the fluid flows, and as such has the advantage of not disrupting the fluid flow. Metallic particles in the fluid passing through the coil will cause small changes in the inductance of the sensor. Ferrous particles will increase the inductance whilst non-ferrous metallic particles will cause a decrease in the flux which is interpreted as a decrease in inductance. A detailed description of such a sensor and the principles involved are given by Whittington (1992).

One of the big advantages of this method is that it can differentiate between ferrous and non-ferrous particles in the on-line mode. However, it does not capture particles for further detailed diagnostic analysis. Another disadvantage is that current implementations can only detect large particles and this implies that this type of sensor may not give very early warning of irregular wear conditions, nor is it suitable for monitoring the general cleanliness of a fluid. It should be noted that the method may experience problems if fluid aeration effects exist. This inductive sensor can be difficult to shield from electrical interference, and can suffer from microphonic problems when experiencing high levels of vibration (Yarrow, 1991).

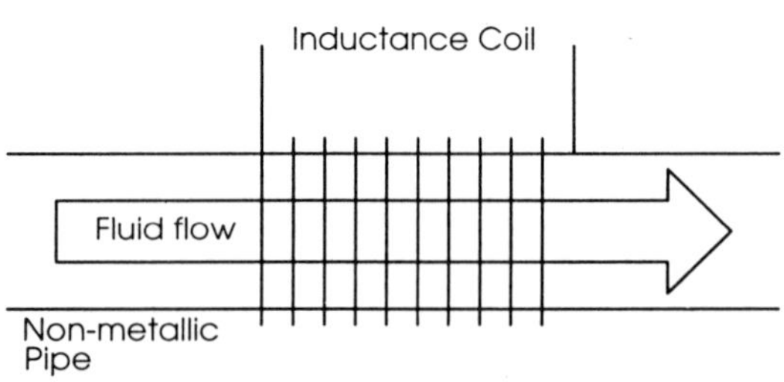

Fig. 5.2 Inductance particle sensor.

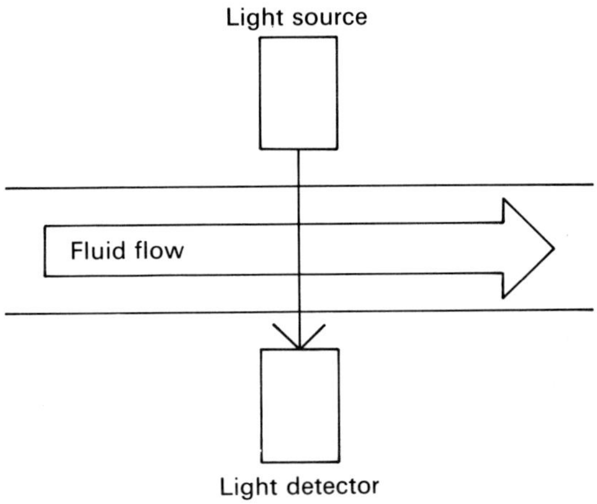

Fig. 5.3 Wear debris detection using light interruption.

5.12 ON-LINE OR OFF-LINE PARTICLE COUNTING USING LIGHT SENSORS

These sensors work on the principle of light interruption and a beam of light is shone across the fluid from a light source. If the light blockage (or extinction) technique is to be employed a light detector is placed on the other size of the fluid so that it experiences a reduction in sensed light intensity as particles pass between the light source and the detector, see Fig. 5.3. If the light scattering principle is employed then light scattering by particles in the fluid is sensed by light detectors.

5.13 ABRASIVITY MONITOR

Until recently, an on-line sensor which measures the abrasivity of non-conducting fluids was produced by Fulmer Systems. The sensor consists of a cylindrical, ceramic substrate on to which have been deposited two thin metallic film resistors with electrical connections to a socket connector. This sensor is located in a housing through which a pressurized flow of fluid is passed. The flow is restricted through a small orifice which projects a narrow stream of fluid on to the wearing film at high velocity. As the film is eroded by abrasive particles in the fluid, so its electrical resistance increases. A reference film, to provide compensation for variations in fluid temperature, is located on the downstream side of the sensor and is shielded from erosion by the fluid. Abrasivity measurements are taken from the sensor by using either a portable hand-held or fixed electronic instrument. An advantage of this monitor is that it responds to ferrous and non-ferrous metallic debris and to non-metallic particles such as ceramic materials and sand. Other advantages are that it is known to be very sensitive and the measurements provided by the method are easy to interpret once the appropriate threshold levels have been established. However, the sensor gives no detailed analysis of the debris – it only detects its presence due to the abrasive action. Another disadvantage is that the rate of erosion will depend upon the hardness and sharpness of the debris and the viscosity of the fluid. Therefore, it is really only ideal where these factors are consistent if reliable trend monitoring is to be achieved.

5.14 FILTER BLOCKAGE DEVICES

Contaminated fluid passing through some form of filter will cause blockage and thereby generate a pressure drop across the filter – see Fig. 5.4. If the fluid flow rate is constant and the number of pores is known then the number of particles can be inferred from the degree of pressure drop. A survey of such techniques is given by Raw and Hunt (1987). The Lindley Fluid Condition Monitor (FCM) (Hunt, 1991) is a commercially available device which uses this principle to measure the particle content in a fluid. The device uses two different size meshes to enable the slope of particle number against size to be assessed. The FCM can be permanently attached to a high pressure line for on-line monitoring via a fluid interface that provides a fixed flow at low pressure. Alternatively, the FCM can be used off-line by connecting its built-in pump to a reservoir or sample container.

The method has the advantage that it is relatively simple and can be implemented on-line or off-line. In the off-line mode it is a relatively quick method of analysis compared to other off-line methods. Other advantages with the method are an ability to deal with fluids which are mixed (oil and water), aerated, heavily contaminated or lightly contaminated. It is

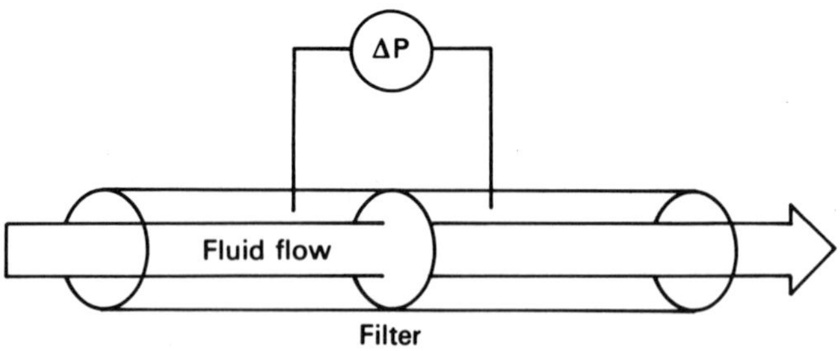

Fig. 5.4 A filter blockage device.

also capable of picking up all types of debris – magnetic or non-magnetic. This makes it ideally suited to cleanliness monitoring and it has proved to be effective in the downstream monitoring of fine filters.

The CEGB developed a sophisticated monitor based on the filter blockage principle (Griffin, 1988). In their implementation the filter membrane is a tape held in a cassette. This tape automatically advances to replace clogged patches when the pressure differential reaches a preset level. The stored filter patches can be used to provide trend analysis, historical records and diagnostics. The monitor operates on-line and employs a bypass pump to draw samples from the monitored system. The samples are then forced through the filter by a slower metering pump. The monitor is capable of detecting particles between one and fifty microns in size. Unfortunately, at the time of going to press the monitor was reported to be unavailable commercially due to a lack of demand.

5.15 COMPARISON OF FLUID ANALYSIS METHODS

The first attribute which must be noted about these methods is the different size of particles that they can detect. Spectroscopy will detect elements in solution but will only detect the presence of particles up to about 5 to 8 microns in size, depending on the specific technique used. Severe mechanical wear produces particles larger than 10 microns, with those larger than 20 microns typically providing a reliable indication of severe wear. Since spectroscopy cannot detect these particles it can result in non-detection of severe wear conditions or, at best, late detection once the debris has become ground down into smaller particles. The ability of ferrography and magnetic plugs to detect the larger particles has lead some authors to argue against the use of spectroscopy for the early detection of severe wear, for example see Lewis (1989); Desjardins and William (1991). Methods for enabling spectroscopy to analyse larger particles have been proposed, for example see Lukas (1991). One such method is to use strong acids to dissolve the large particles but this can be time consuming and greatly increases the cost of the relatively cheap spectrographic process. Furthermore, Desjardins and William (1991) argue that even if the larger wear debris is dissolved its effect on the overall iron concentration levels in the sample fluid will be very small, i.e. iron concentration monitoring is not the same as large particle detection.

Nevertheless, spectroscopy is still a very popular technique. It has the advantage over ferrography and magnetic plugs that it can detect a wide range of magnetic and non-magnetic elements, as described above, and it can detect elements in solution as well as particle form. These features make spectroscopy particularly useful for monitoring oil additives and the ingress of contaminants such as siliceous material from the atmosphere and general airborne dust. The more comprehensive analysis offered by spectroscopy may give information which can pinpoint the exact location of an impending failure. If the metallurgy of the various components of the monitored system is known, then the ratio of the elements in the wear debris may indicate the source of wear, i.e. a form of signature analysis can be performed using signatures which define the quantity of elements such as iron, chromium, nickel, molybdenum and magnesium. Another characteristic of spectroscopy is that it is an extremely sensitive technique which again suits it to cleanliness monitoring.

It may well be that in a particular application spectroscopy will be used in addition to one of the other techniques, such as magnetic plugs, the other technique being employed to give an early warning of severe wear conditions, possibly in an on-line mode. Evans (1991) argues the case for spectrometric oil analysis being a complement to ferrographic analysis.

Ferrography will detect particles from about 1 to 100 microns in size. Magnetic plugs on the other hand will pick up particles from about 20 to 30 microns in size to beyond 1000 microns. This means that magnetic plugs are more advantageous when monitoring ferrous components which produce wear debris in large pieces, e.g. rolling element bearings and gears. It should be remembered that large particles may not be picked up in oil samples since they fall to the bottom rather than remaining in suspension.

The rotary particle depositor has been developed to overcome conventional ferrography's inability to deal with large sample volumes. Conventional ferrography techniques can normally only cope with a sample volume of about 1 ml. Larger sample volumes will obviously give an analysis with tighter confidence limits and they are less likely to exclude large pieces of wear debris which occur in small numbers.

In general, the currently available on-line techniques do not give detailed fluid analysis and fault diagnosis information. Instead, they are used to detect abnormally high concentrations of wear debris or contamination and large pieces of wear debris. On-line sensors can be used downstream from filters to monitor the effectiveness of the filtration system and the cleanliness of fluid. One important advantage of on-line systems, from the practical 'shop-floor' point of view, is that they are not dependent on oil sampling which, as described above, must be carried out with the utmost care.

5.16 FLUID ANALYSIS VERSUS VIBRATION MONITORING

One of the key advantages of fluid analysis is that, by monitoring contamination and fluid condition, it can detect the conditions which will cause severe wear, or malfunctions, before such effects have taken hold. From a practical point of view, it can be argued that vibration monitoring is easier to implement in terms of fundamental signal acquisition. Also, vibration monitoring lends itself more readily to on-line (real time) monitoring.

However, the condition monitoring industry is increasingly taking the view that fluid and vibration analysis should be seen as complementary techniques rather than competitors. Hunt (1993) cites several references which endorse this view in the light of practical experience. The use of both techniques can provide wider fault coverage and corroboratory information when making critical decisions. Having the skills and facilities required to apply both techniques can enable users to cater with a much wider range of monitoring situations since in some situations only one of the techniques may be applicable. For example, slow machinery may be more effectively monitored using fluid analysis, whilst the source of wear debris in a series of hydraulic pumps, connected to the same fluid supply, cannot be pinpointed by fluid analysis.

5.17 REFERENCES

Albidewi, A. *et al.* (1991) Determination of particle shape by image analysis – the basis for developing an expert system, in *Proceedings of Condition Monitoring '91*, Erding, Germany, 14–16 May, 1991, 411–422.

Anderson, D.P. (1986) *Wear Particle Atlas* (Revised), Naval Engineering Center, Lakehurst, New Jersey, USA, Report NAEC-92-163, 26 June, 1986.

Davies, A. (ed.) (1990) *Management Guide to Condition Monitoring in Manufacture*, Institution of Production Engineers.

Desjardins, J.B. and William, W.S. (1991) Comparative effectiveness of measuring iron particles using spectrometric and ferrographic techniques, in *Proceedings of Condition Monitoring '91*, Erding, Germany, 14–16 May, 1991, 364–371.

DTI (1980) *Contamination Control in Fluid Power Systems Vol. 1 – Field Studies*, Department of Trade and Industry, NEL, East Kilbride, Glasgow, UK.

Evans, W.A. (1991) Spectrometric oil analysis: a complement to ferrographic analysis. *Lubrication Engineering*, **47** 6, June 1991, 437–439.

Griffin, R. (1988) Continuous wear debris monitoring. *CME*, April 1988.

Hunt, T.M. (1991) Debris analysis on-line and off-line – the filter blockage technique at work, in *Proceedings of Condition Monitoring '91*, Erding, Germany, 14–16 May, 1991, 339–349.

Hunt, T.M. (1993) Vibration and wear debris analysis – how they correlate and complement each other, in *Proceedings of 5th International Congress on Condition Monitoring and Diagnostic Engineering Management*, UWE, Bristol, 21–23 July, 1993, 109–114.

Jones, M.H. (1989) Wear debris analysis of grease, in *1989 Condition Monitoring and Preventative Maintenance Conference Proceedings*, Atlanta, Georgia, May 1–4, 1989, 132–139.

Lewis, R.T. (1989) The mechanical component of wear, what spectroscopy doesn't tell you, in *1989 Condition Monitoring and Preventative Maintenance Conference Proceedings*, Atlanta, Georgia, May 1–4, 1989, 78–82.

Lukas, M. and Anderson, D.P. (1991) Techniques to improve the ability of spectroscopy to detect large wear particles in lubricating oils, in *Proceedings of Condition Monitoring '91*, Erding, Germany, 14–16 May, 1991, 372–398.

National Coal Board (1984) *Wear Particle Atlas*, Headquarters Technical Department, NCB, Bretby, UK.

Price, A.L. and Roylance, B.J. (1984) The Rotary Particle Depositor – a response to problems experienced with wear particle deposition, in *Proceedings of Condition Monitoring '84*, Swansea, 1984, 596–607.

Raw, I. and Hunt, T.M. (1987) *A Particle Size Analyser based on Filter Blockage*, Condition Monitoring '87, International Conference, Swansea, 875–894.

Uedelhoven, W., *et al.* (1991) The use of automated image analysis for the study of wear particles in oil lubricated tribological systems, in *Proceedings of Condition Monitoring '91*, Erding, Germany, 14–16 May, 1991, 423–435.

Veinot, D.E. and Fisher, G.C. (1989) Wear debris as a condition monitoring technique for the Sikorsky Sea King Helicopter main gearbox, in *1989 Condition Monitoring and Preventative Maintenance Conference Proceedings*, Atlanta, Georgia, May 1–4, 1989, 119–131.

Whittington, H.W. and Flynn, B.W. (1992) Wear debris monitoring in lubricant. *Condition Monitoring and Diagnostic Technology*, **3** 2, November 1992, 61–64.

Yarrow, A.S. (1991) Condition monitoring by wear debris analysis. *Noise and Vibration Worldwide*, **22** 5, May 1991, 24–28.

Systems approach to condition monitoring

6

6.1 INTRODUCTION

This chapter considers the systems approach to condition monitoring (CM). Traditionally characteristics of components that need to be monitored such as bearings, seals, ballscrews etc., are measured by sensors on a one-to-one basis. These components form sub-assemblies which in turn form the complete system. It would be more efficient if these component characteristics could be monitored by sensors at the sub-assembly or even complete system level.

Systems engineering has become the conventional approach in the manufacturing domain (Williams and Davies, 1992). This chapter considers those aspects of a system formulation which are appropriate to CM. In this application only systems that are characterized by their dynamic behaviour are suitable. Further, the systems approach dictates the manner of CM. Use is made of diagnostic techniques that have been developed for fault location of complex dynamic systems.

6.2 ONE-TO-ONE CONDITION MONITORING

One-to-one condition monitoring means monitoring the value of a parameter that is a measure of some particular characteristic of a sub-assembly with a suitable sensor. This is a one-to-one configuration in that one sensor relates to one characteristic as illustrated in Fig. 6.1.

Subsequently, using a one-to-one arrangement for many physical characteristics will result in the need to establish a data processing infrastructure to give meaningful information for operational control. This is clearly inefficient but a natural consequence if a sensor has to be added

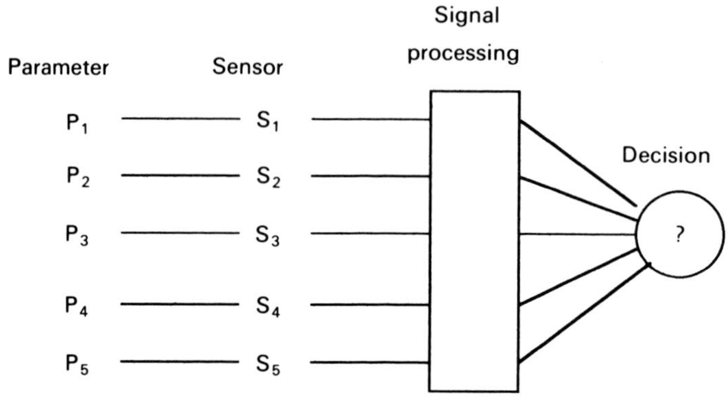

Fig. 6.1 One-to-one condition monitoring.

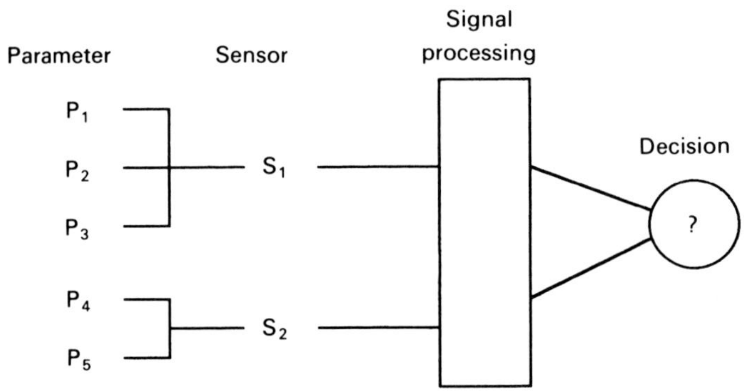

Fig. 6.2 One-to-many condition monitoring.

to measure each characteristic. A good example of the one-to-one application is the monitoring of a bearing using an accelerometer attached to the bearing housing that enables the vibration level to be measured, see Chapter 4.

In one-to-many CM the output of one sensor is used to give CM information on more than one sub-assembly, as illustrated in Fig. 6.2. If achievable this is clearly more efficient. As a consequence, the sensor used is not chosen to monitor a particular physical characteristic but rather to produce a signal that can be analysed to give information in a parametric sense on a number of physical characteristics. In fact it is precisely in this situation that many techniques exist in the general field of system fault location. Fig. 6.3 illustrates how ease of access to components diminishes as one moves from component to sub-assembly to complete system.

So monitoring at the sub-assembly level, while naturally reducing the number of sensors, can make it more difficult to get information on

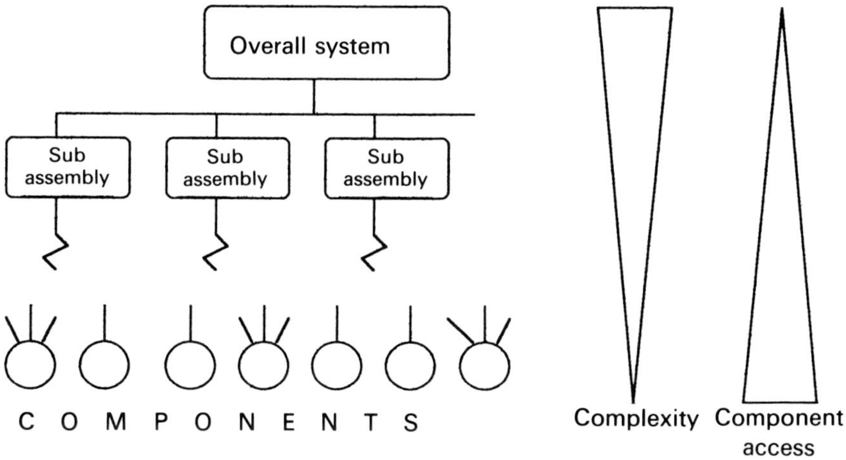

Fig. 6.3 Complexity versus availability of component information.

components. Using one data source to gain information on more than one component is a feature of system-based condition monitoring. Basic features of system configuration and performance need to be considered before system condition monitoring techniques are developed.

6.3 SYSTEM REPRESENTATION

> 'system: anything formed of parts placed together or adjusted into a regular and connected whole.'
>
> *Chambers (1983)*

A simple switch can be designated as a system 'parts··· together··· connected whole'. However, in operation, it is perceived sensibly as a single entity having binary states of on or off. Thus the dictionary definition of a system needs to be qualified for use in condition monitoring. What really confirms the qualitative feel of a system is its dynamic behaviour. For example, a jigsaw puzzle meets the definition exactly as does a robotic arm. While one would describe the arm as a system one would not so describe the jigsaw puzzle. The principal difference is that the robotic arm is characterized by a differential equation which describes its dynamic behaviour while the jigsaw puzzle is not. In practice the dynamics of a system are conveniently represented by the system's response to a known input stimulus. Diagrammatically the relationship is shown in Fig. 6.4. This is one of the possible number of ways of basic system representation showing that it receives a demand and responds, dynamically, to produce

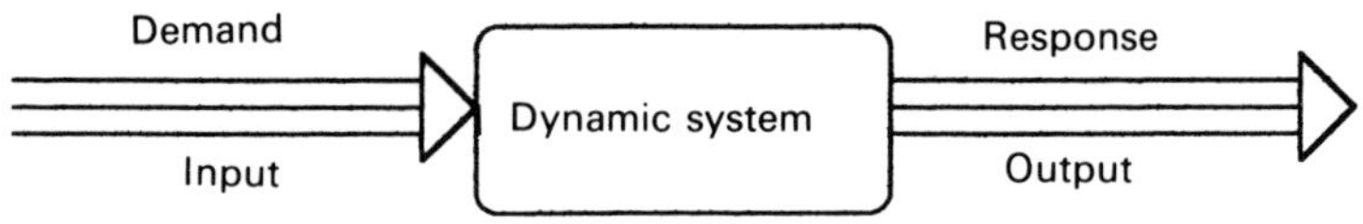

Fig. 6.4 Basic system representation.

a response (Dorf, 1980). The terms 'input' and 'output' need some clarification. These terms are pertinent for the diagrammatic representation. Input defines the externally generated signals or data that the system receives. Output is any convenient measure of the change in the system's state or states. It does not have to be the physical output of the system.

Dynamic behaviour is described using differential equations. It is not intended to give a comprehensive description of dynamic models but to select models and techniques that are adequate for condition monitoring. All models are 'man-made' and are simply convenient representations of actual devices.

6.4 MATHEMATICAL DESCRIPTION

For completeness the basic mathematical system representation as used for synthesis and analysis will be presented. It is assumed that the models are linear time invariant. First we will examine the similarity between the dynamic descriptions of mechanical and electrical systems. This similarity enables general rather than particular models to be used which results in a unified approach to systems based CM. Consider the two systems shown in Fig. 6.5, one mechanical, one electrical. Using Newton's Laws of Motion for the mechanical system gives:

$$M\frac{d^2y(t)}{dt^2} + f\frac{dy(t)}{dt} + Ry(t) = r(t) \tag{6.1}$$

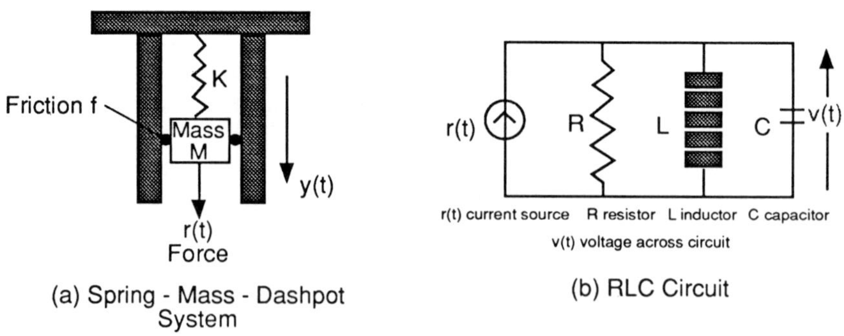

Fig. 6.5 Mechanical and electrical systems.

Here the first term is the force due to the acceleration of the mass M; the second term is the force due to f, the frictional force which is dependent on the velocity; and the third term is the force due to the spring's extension. These forces must balance the applied force r(t). The method of solution (Dorf, 1980) is beyond the scope of this book but if the mass is initially displaced then released the solution is of the form:

$$y(t) = K_1 e^{-\alpha t} \sin(t + \beta) \tag{6.2}$$

Using Kirchhoff's current law the RLC circuit can be described as:

$$C\frac{dv(t)}{dt} + \frac{1}{R}v(t) + \frac{1}{L}\int_0^t v(t)dt = r(t) \tag{6.3}$$

The terms on the left-hand side are the three currents through the capacitor, resistor and inductor respectively. These currents must sum to the applied current, r(t). If a constant current is applied the solution has the form:

$$v(t) = K_2 e^{-\alpha t} \sin(t + \beta) \tag{6.4}$$

It is seen that both systems have equivalent responses. Further if equation 6.1 is rewritten in terms of velocity rather than displacement we have:

$$M\frac{dv(t)}{dt} + fv(t) + R\int_0^t v(t)dt = r(t) \tag{6.5}$$

It is clear that the two systems are analogous. This is the attraction of a system's approach where many diverse systems are in fact realizations of the same underlying dynamic system. The three degrees of freedom are the gain, which is a scaling factor; the damping ratio, which if greater than unity produces an over damped response and conversely an under damped system; and the undamped natural frequency which governs the speed of response. Linear, time-invariant models are used because they are solvable. Non-linear representations may be more realistic but are difficult or even impossible to solve analytically. Superficially, it seems natural to develop as comprehensive a model as possible. This means that to take account of all the dynamic features and nuances of a given physical system would result in a high-order model. For instance, the frictional term in equation 6.1 is idealized in that the frictional force is assumed to be only linearly dependent on the velocity of the mass. A model taking into account all the non-linearities may well give an accurate representation of the performance of the physical system. However, in no sense would it be a generic model of the system type for such a model would be sensitive to the inherent small variations between a number of realizations of a particular artefact. The modelling process would have to be repeated for every system of the same type.

The point is, that no model will completely define the system so why not use one that is easily established and analysed? What is critical is that the chosen model should describe the **fundamental** dynamic behaviour. This implies the model should be as simple, or low order, as possible while being realistic in a broad sense (Williams, 1985).

6.5 USEFUL LOW ORDER MODELS

The simplest model is the first order model with one degree of freedom but used alone its use is usually restricted to the representation of a simple delay. This is too much of a restriction for CM applications. Probably the most useful model 'building-block' is a second order model as already considered. The model's dynamics can be conveniently presented by giving the response of the system to a given input. It does not mean that the system will actually suffer such an input but rather, the response to such an input is a convenient dynamic description. Often this input/output relationship is presented graphically rather than with mathematical expressions as in equations 6.2. and 6.4. To do this one has to first decide what would be a suitable system input. The fundamental input is the unit impulse. The definition of an impulse is an infinite 'spike' lasting for an infinitesimally small time interval. While this concept is a useful device for the mathematical analysis of the system it is obviously not a practical choice for an input. A more realistic input is a change in the input level; a step change, i.e. displacing the mass and letting go or applying a steady current to the RLC circuit. The model's response depends upon the magnitude of the step change and the three parameters of the model (a

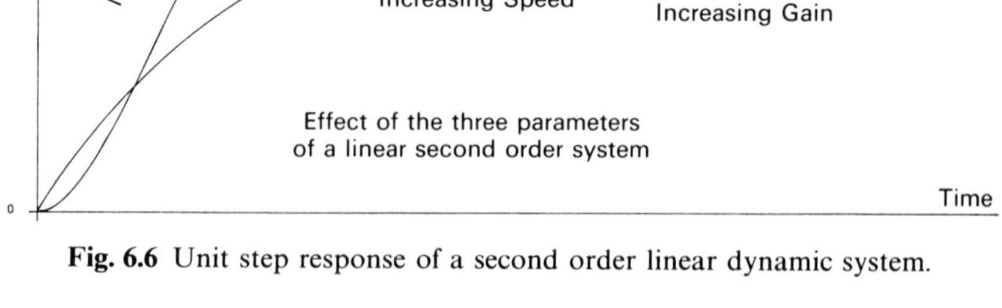

Fig. 6.6 Unit step response of a second order linear dynamic system.

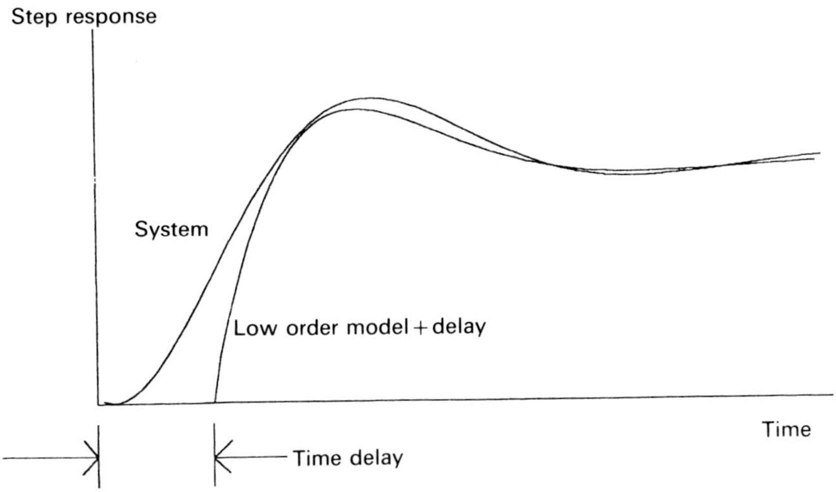

Fig. 6.7 Use of low order model for system representation.

second order model has three parameters; for example, M, f and K in the mechanical system and R, L and C in the electrical circuit).

Figure 6.6 shows the two possible types of response to a step input that a second order system can produce, under damped and over damped. There is a third which is the boundary between the two, namely critically damped (Dorf, 1980). In reality this 'knife-edge' response is effectively unattainable. The systems parameters are the gain, which is a scaling factor; the damping ratio, which if greater than unity produces an over damped response and conversely an under damped system; the undamped natural frequency which governs the speed of response. An additional degree of freedom is obtained by shifting the response, or delaying, along the time axis. This technique is useful for systems that would need much higher order models than second order, for the extra terms affect the response near the origin. Thus by shifting along the time axis the initial conditions are 'ignored'. Figure 6.7 shows the basic idea. The usefulness of the second order model is that with parameter values it can represent the dynamic behaviour of a wide range of systems.

6.6 GENERIC MODELS USING THE LAPLACE TRANSFORM

The mathematical representations and solutions as developed in equations 6.1 to 6.4, while accurate under linear constraints, are not convenient to use for general classification and use. System analysis utilizes techniques of the Laplace transform (Dorf, 1980). Using a Laplace representation and restricting ourselves to systems that are assumed to be 'at rest' initially effectively removes the need to use the often complex mathematical treatment of differential equations. First we will consider the transformation of the differential equation of the system.

(i) v(t) transforms to v(s). This simply means that the time varying variable v(t) is represented in the Laplace domain as v(s).
(ii) dv(t)/dt transforms to sv(s). Basically the derivative d/dt becomes s. This is true if v(t) and all its derivatives are zero at time $t = 0$.
(iii) In general the *n*th derivative becomes the *n*th power of s. Thus a general linear differential equation is readily transformed into the Laplace domain by using these simple rules term by term. Consider an *n*th order differential equation:

$$a_n \frac{d^n y(t)}{dt^n} + a_{n-1} \frac{d^{n-1} y(t)}{dt^{n-1}} + \cdots + a_1 \frac{dy(t)}{dt} + a_0 = x(t)$$

Then transforming term by term yields:

$$[a_n s^n + a_{n-1} s^{n-1} + \cdots\cdots + a_1 s + a_0] y(s) = x(s)$$

or

$$\frac{y}{x}(s) = \frac{1}{a_n s^n + a_{n-1} s^{n-1} + \cdots + a_1 s + a_0}$$

This ratio of output, y(s), to input, x(s), is called the transfer function of the system, G(s) say. So to determine the response of the system to any given input, z(t), one transforms z(t) to z(s) and 'operates' on z(s) with G(s).

$$z(t) \rightarrow z(s)$$
$$\text{and } y(s) = G(s)z(s)$$

This is the Laplace transform of the response and to obtain the actual response, y(t), we have to determine the inverse Laplace response of G(s).z(s).

So far we have considered the transformation of derivatives, which given zero initial conditions, is seen to be straightforward. The transformation and inverse transformations of other functions are generally restricted to functions that are useful in engineering. Of course the transformation can be applied to any time dependent variable but generally we are only concerned with functions useful from an engineering viewpoint. Examples of Laplace pairs are given in Table 6.1. The main advantage of considering systems that are represented by linear time invariant differential equations is that the response is dominated by exponential functions as seen in equation 6.2.

Table 6.1 Useful Laplace transform pairs

Function	*Time plot*	*Laplace transform*
unit step $f(t) = 0, t < 0$ $= 1, t \geqslant 0$	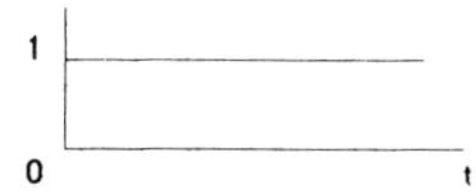	$f(s) = 1/s$
unit ramp $f(t) = 0, t < 0$ $= t, t \geqslant 0$	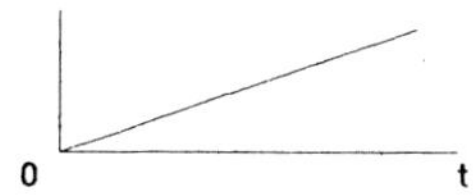	$f(s) = 1/s^2$
exponential $y(t) = e^{-at}$	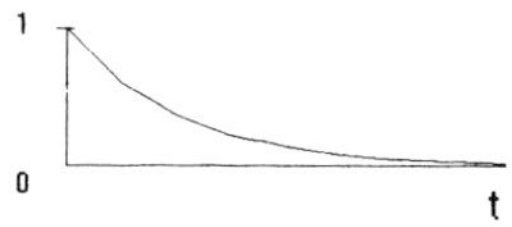	$f(s) = 1/(s + a)$
sine wave $f(t) = \sin(\omega t)$	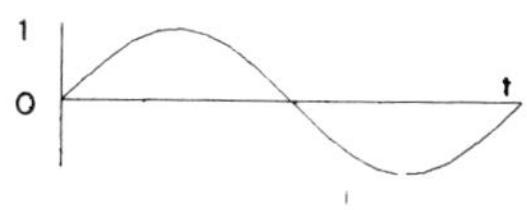	$f(s) = \omega/(s^2 + \omega^2)$
cosine wave $f(t) = \cos(\omega t)$	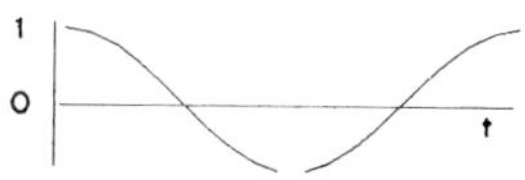	$f(s) = s/(s^2 + \omega^2)$
integral $z(t) = \int f(t)dt$		$z(s) = f(s)/s$

6.7 CONFIGURATION REDUCTION

So far we have considered the system as a single block with an input and an output. Practical systems are in general a collection of such blocks with various input/output connections. Figure 6.8 shows the block diagram of an automated guided vehicle (AGV) control system. This is simply a representation of the actual system. It is convenient as the major sub-assemblies of the whole system are clearly identified such as transducer signals (position and speed), electronic systems (power amplifier) and electro-mechanical systems (brake system) etc. It must be stressed however that it is only a convenient representation and the boundaries between the various subsystems are not in general so well defined. It is seen that the AGV system includes feedback connections of the vehicle's speed and position. Feedback is found in all dynamic systems of interest. The use of feedback reduces variation in system performance due to degradation in the system's components and external disturbances. This is the main advantage of such a closed-loop system over the system without feedback,

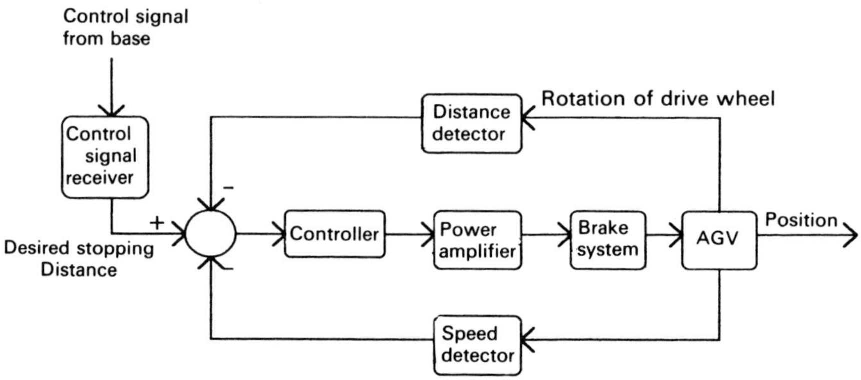

Fig. 6.8 AGV control system.

called an open-loop system as shown in Fig. 6.4. However there are penalties. Open-loop systems are inherently stable whilst feedback can cause instabilities. Stability analysis is of primary concern to system designers. A closed-loop system uses more components and is more complex than an open-loop one. In addition there is a loss of gain in using feedback but this loss is 'used' to reduce the effects due to component variation and disturbances.

All systems such as the AGV systems can be reduced to a single block with an input and output. This is achieved using straightforward algebraic rules. These are summarized in Fig. 6.9.

This reduction produces a relationship or transfer between two signals of interest. For example suppose the relationship between the 'desired position' and the 'actual position' was required. Using algebraic rules the

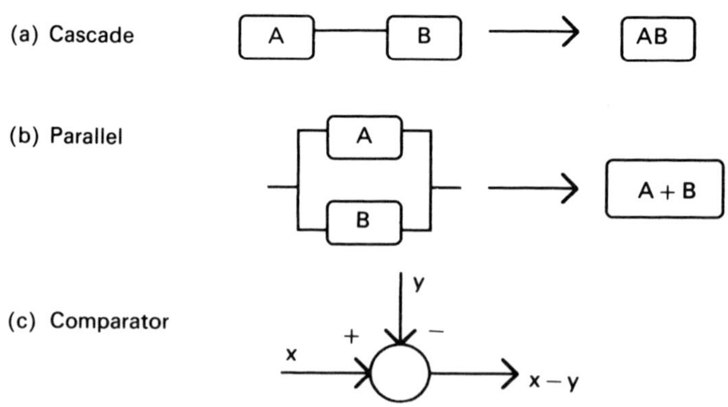

Fig. 6.9 Basic rules for configuration reduction.

multi-block representation can be reduced to a single block, as shown in Fig. 6.4. Of course the dynamic equation in this single block is of a much higher order than any one of the individual blocks of the original. Although complexity has been concentrated into a single block the use of a low order approximation of the resulting high order model can enable the performance of the overall system to be readily estimated. This reduction relates desired position to actual position. Whilst this is a sensible selection, any two signals can be chosen as the input and output of a single block, e.g. the input could have been 'control signal from base' and the output 'rotation of the drive wheel'. Of course these choices would have resulted in a different transfer function. The point is that for any system one is free to choose convenient inputs and outputs. This is particularly important in CM where access to signals is often difficult. Not enough thought is given in system design to maintenance requirements and one is forced to use available access points.

6.8 PRACTICAL MODELLING

So far we have seen how a Laplace-based system model can be utilized to represent the dynamics of the sub-assembly of interest. The use of low order models has been stressed. The usefulness of low order models can be extended by recognising the fact that an error will exist between the model's response and the actual response of the system. Figure 6.10 shows the basic idea.

As time increases from zero the error will grow unless the model is re-initiated. The intervals between re-initiations can be used as a measure

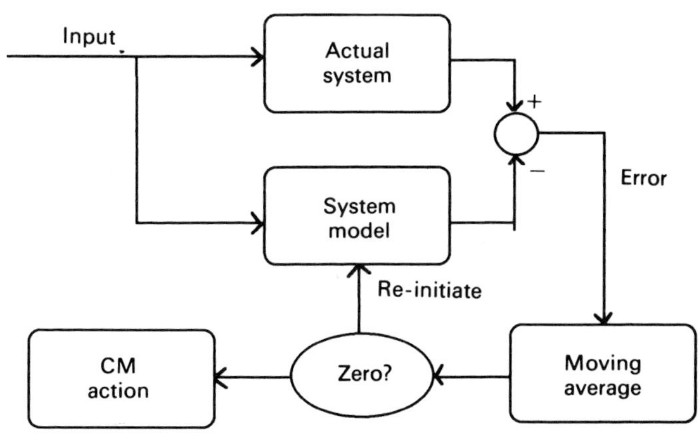

Fig. 6.10 Error rate based CM.

(a) Model known (b) Model unknown

Fig. 6.11 Input model or input/output system representation.

of the condition of the system. The intervals are averaged over the time and if the degree of match between the model and the actual system is unchanged then this moving average will be effectively constant. If the dynamics of the actual system changes then the average time between re-initiations will reduce. Thus we have a means of incorporating the natural inaccuracy of a low order model into a useful source of CM information.

The input/output relationship can be used without detailed knowledge of the 'system' between these two access points. The input and output signals can be created as a dual characteristic of the system. The establishment of the system model enables the output to be estimated for any input. However in practice systems only suffer a known subset of possible demands. For example, the table of a machine tool will regularly move from a datum to a given location prior to machining. So the input/output signals during this motion can be used as a dynamic description of the system as an alternative to assuming a model and predicting its response, as illustrated in Fig. 6.11. A convenient approach is to simply monitor the designated output for a known input. So for the machine tool example, the output is noted whilst the table moves from the datum.

This signal is then the characteristic dynamic 'fingerprint' of the system. CM is then implemented by observing changes in this output signal under repeatable conditions.

6.9 CM DIAGNOSTICS

Inherent in CM is that when the system is deemed to have moved into a state where maintenance action has to take place, it must be determined what component, sub-assembly or parameter setting is at fault, see Fig. 6.12. This is then diagnostics. CM is a two stage procedure, first checkout then if required, diagnostics. If one adds the need to predict the likelihood of future failure then diagnostics are needed even if the system passes the checkout stage.

An operating process moves from a state of 'correct' working to a state of 'incorrect' working. The mode of translation from one state to the other

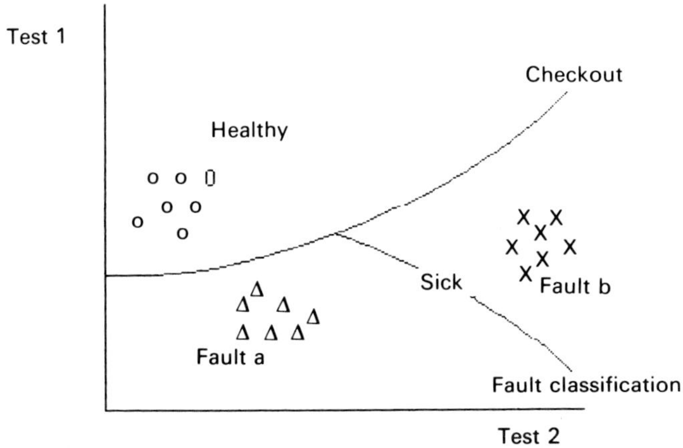

Fig. 6.12 Illustration of fault classification.

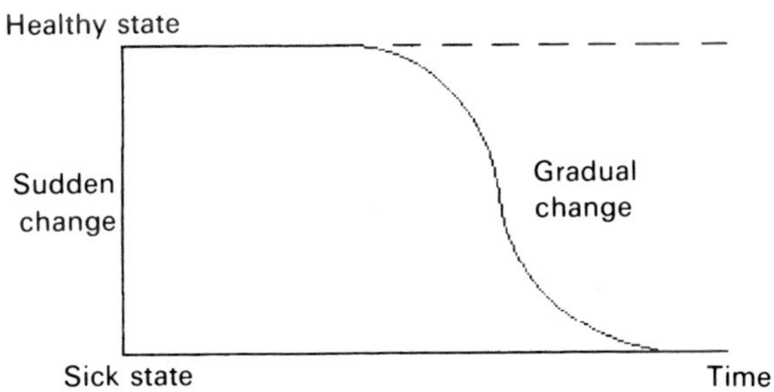

Fig. 6.13 Modes of failure.

depends upon the particular cause of 'failure', see Fig. 6.13. The terminology used is often of a medical source. One describes a system as 'healthy' or 'sick'. This is a useful analogy to make, for it enables the relationship between CM and diagnostics to be easily made. Diagnostics are instigated when the system is sick. There are two accepted modes in CM. One when the system suddenly moves from healthy to sick (catastrophic) and the other when the system is monitored on the gradual path from healthy to sick (drift).

The more challenging CM is concerned with the second change of state, namely the drift fault. Using the medical analogy, the system is always 'off-colour'. It is the degree of 'off-colour' that determines if any remedial action is required. It is predictive in that the system is observed periodically and checked against criteria that establish whether the system is healthy enough for continued operation, see Fig. 6.14.

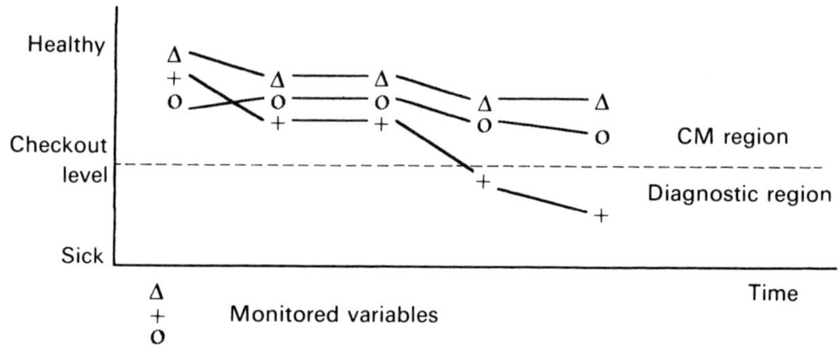

Fig. 6.14 Relationship between CM and diagnostics.

It is advantageous to utilize the diagnostic techniques not only in the diagnostic region but in the healthy region as well. This will enable the state of health of the various sub-assemblies to be established giving prior warning of a likely failure, and enabling management to initiate planned maintenance.

6.10 SYSTEM FAULT LOCATION TECHNIQUES

This is not the place to give an exhaustive survey of the techniques (Williams, 1985) but the key factors will be addressed. Detailed considera-

Table 6.2 Classification of analogue fault location methods

Sequential methods		Element by element Half split method Dynamic programming
Non-sequential methods	Parameter estimation	Interactive method Model adjustment Parameter adjustment Parameter updating Connectivity functions
	Probability methods	Key element search Inverse probability Entropy method
	Pattern recognition	Matrix recognition Sequential recognition Bode diagram Bilinear transform Cross production method Voting technique Template matching Nearest neighbour Learning technique

tion is given in Chapter 3. Here we will highlight those that are useful in a system sense.

Where possible any diagnostic technique should use normal operating records. Under CM the diagnostic routines should not interrupt normal working. One possibility is to exercise the system when in its idle state. However, this may make the overall operation too complicated. Further, it is better to have access to information while the process is operating. The dynamics of the device are monitored rather than particular static measures. Static tolerances have to be much tighter than those set for dynamic behaviour and most systems are designed to give a particular dynamic performance (Williams, 1985). Many systems are designed to respond to real time varying input signals, so that a high degree of correlation exists between operational success and the setting of suitable dynamic performance specifications. Sequential and non-sequential diagnostic methods are shown in Table 6.2. This is a general list and not all the techniques are suitable for CM.

6.11 DISCRETE AND CONTINUOUS DIAGNOSTICS

System diagnostics can be conveniently classified into two classes, discrete and continuous. Whilst analogous to the descriptions of signals, here they describe the types of failure modes. In diagnostic terms they refer to hard and soft faults. The basic differences are listed as:

- Hard
  (a) large change from specification
  (b) often change in system structure
  (c) catastrophic
- Soft
  (a) drift from specification
  (b) system structure unaltered
  (c) slow change.

Hard fault diagnostics itself can be further subdivided into two further classes. The state of the device can take one of two such as on or off, short or open circuit, i.e. binary states.

A switch is such a device. There exist techniques such as fault tree analysis that can be used to locate such faults in binary elements of a system. Being of a binary nature the diagnostics utilizes logic analysis and well established Boolean optimization.

Alternatively, the device can be designated as being in one of three discrete states such as nominal, high or low. This quantization can be used on a signal that can drift continuously over a wide range such as

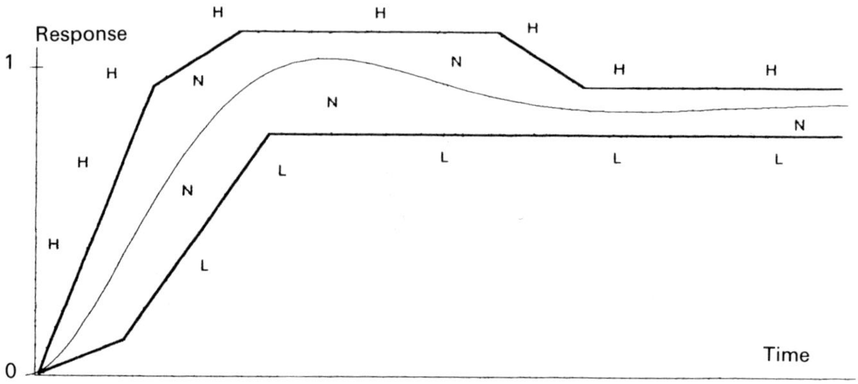

Fig. 6.15 Quantization using response template.

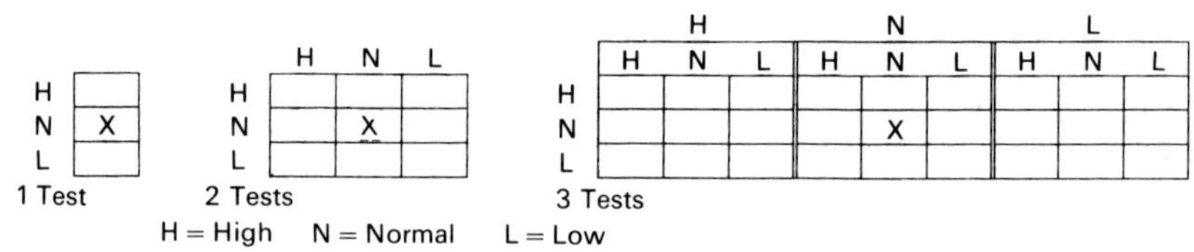

Fig. 6.16 Fault cover as a function of number of tests.

the output from an amplifier. The method is to establish a tolerance band around the chosen response which results in three regions being established, within, above and below tolerance.

The template is relative to a continuous signal. To use this information sample times must be chosen from the plot. Figure 6.15 shows a typical template for checkout using the step response which illustrates the quantization levels.

Figure 6.16 shows how the cover increases with number of sample points. The number of such combinations is equal to $n^3 - 1$, where n is the number of tests. Of course this is the greatest number of possible faults that can be covered, not necessarily the actual number achievable for a given system. This depends upon the discriminatory power of the tests. This will be addressed later.

6.12 RECOGNITION MATRIX

Each likely fault will affect the system's response producing a deviation vector that is characteristic of the fault. If the nominal response is subtracted

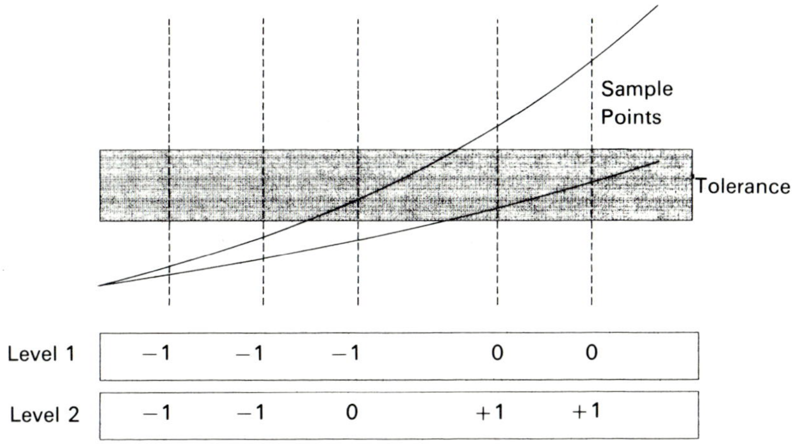

Fig. 6.17 Change in signature due to different fault levels.

from this fault response we will obtain a deviation vector that 'represents' the particular fault. If this deviation vector is then compared to the tolerance template at selected times then a pattern of ± 1 s and 0 s can be established, see Fig. 6.17. This quantized vector is a 'fingerprint' or signature of the fault. This procedure is repeated for all the likely faults. This results in a set of signatures that collectively is the recognition matrix, sometimes called the fault dictionary, of the system. This vector set, the recognition matrix, is the 'memory' of the faults.

Having established the recognition matrix a diagnostic scheme can be implemented using this 'memory'. For a system that has failed the checkout stage the deviation vector is determined by subtracting the nominal response. This is then quantized into the three levels producing a fault signature. This is then compared to the fault signatures of the recognition matrix and the fault identified by matching the fault signature with one of the stored signatures of a known fault.

6.13 SIGNATURE MATCHING

There are various methods of matching quantized signatures. One is to simply multiply the two vectors:

$$(\mathbf{x}, \mathbf{y}) = [-1, -1, 0, 0, 1, 1][-1, -1, -1, 0, 0, 1]^{\mathrm{T}} = 3$$

While this is an easy procedure to automate, information is lost. The problem is that 0 produces 0 in a product. An alternative is to match element by element and score 1 for each match. In this case the above example would produce a total of 4.

So far we have assumed ideal noise-free signals. In practice this is rarely realized. In addition the nominal response assumes that all components and/or parameters are at their nominal values. This never occurs. The expected variations in component values give rise to the tolerance band of the system's response, as shown in Fig. 6.15. This variation will be present in the response under fault conditions, although in this case the variation will be dominated by signal noise. Quantized methods go some way in negating this problem in that the quantized bands will allow some variation from their expected levels. This together with its simplicity makes the quantized pattern matching method an attractive approach. However the method suffers from two major drawbacks. If the fault can take on a wide range of possible levels rather than the three bands of high, normal and low then the three level quantization is inadequate. Figure 6.17 shows how the fault signature can change when the faulty component can take on different fault levels. This means that numerous signatures are needed to represent a single fault. In this situation confusion can arise between two or more faults that can give the same quantized signatures whilst experiencing differing fault levels. In other words the method is based on the magnitudes of the fault deviations.

The other drawback is that quantization excludes some of the information contained in the deviation vector. It would be better if the deviations were stored more precisely. (An example of the loss of information by quantization is if π is given a value of 3.) These drawbacks have led to the development of other techniques that are based on the shape of the fault deviations rather than the magnitude of the deviations.

6.14 NON-QUANTIZED METHODS

It's the **shape** of the deviation vector rather than the magnitude at the sample points that really represents or characterizes the fault. Figure 6.18 shows how the three sample values can be represented by a vector. Of course there can be more than three sample points in practice but by considering only three the ideas can be represented graphically. The diagram shows that if the level of the fault increases the **magnitude** of the vector can change much more than the **direction** of the vector. The situation can be represented by a 'fault-cone', see Fig. 6.19, within which the vectors representing a particular fault lie. So it would be advantageous if the fault deviations were represented by the direction of the vector rather than its magnitude. This can be achieved by normalizing the vector. Of course more than three coordinate axes are difficult to visualize but the

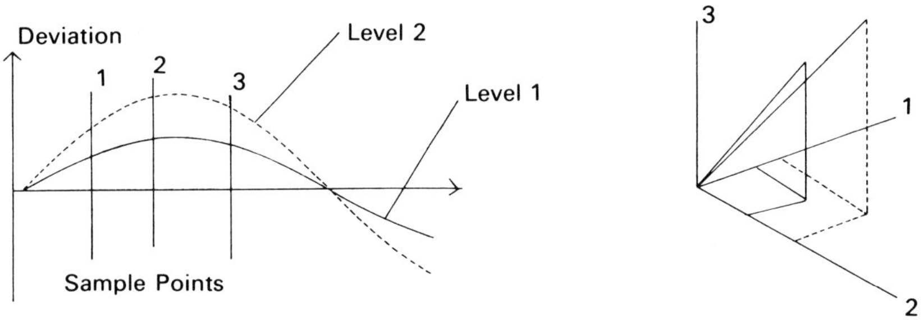

Fig. 6.18 Fault deviation represented by vectors.

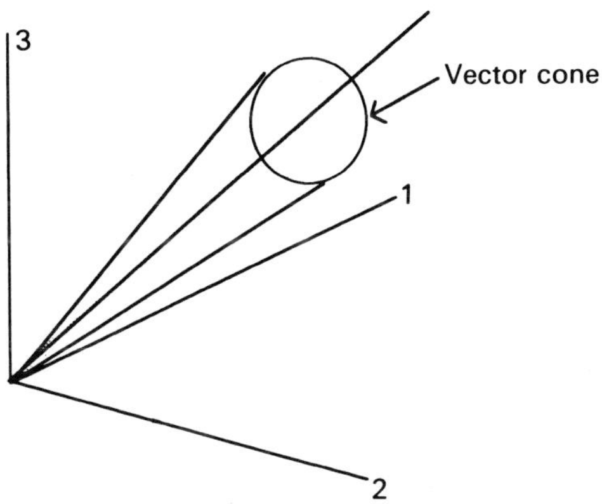

Fig. 6.19 Fault vector cone.

mathematical measures such as length and direction are applicable to vectors of more than three dimensions.

Having established a normalizing method that eliminates to a large degree the effect of varying fault levels we now need a method of matching the fault vector of the system under test with the stored vectors in the fault dictionary. A well known method is the 'nearest neighbour'.

The nearest neighbour technique uses the distance between vectors as a measure of their 'nearest'. The smallest value of the distances between the fault vector and the stored vectors is taken to indicate the most likely fault. In fact these calculated distances can be ranked in ascending order and thus the corresponding faults ranked in the order of likelihood. This nearest neighbour method is the most useful technique in CM associated with parameter values that are subject to drift. It is a method that is easily understood and automated.

6.15 CONCLUSION

This chapter has been a tour of the features of system based condition monitoring. The fundamental characteristic is the opportunity of obtaining CM information of many parameters from a single data source. Classic system representation uses the Laplace transform representation of the differential equations describing the dynamic system under consideration. Practical CM does not need the analytic tools used in dynamic analysis but rather the concepts of the systems approach of input/output relationships.

Methods of diagnostics fall into two classes, quantized and non-quantized. Quantized methods have echoes with fault tree analysis where the components being monitored can take only one of two states. Non-quantized methods are applicable in situations where component values can drift with time and can be used in a predictive sense, as considered in Chapter 8.

6.16 REFERENCES

Dorf, R.C. (1980) *Modern Control Systems* (2nd ed.), Addison-Wesley, Reading, MA.

Williams, J.H. (1985) *Transfer Function Techniques and Fault Location*, Research Studies Press, Wiley & Sons.

Williams, J.H. and Davies, A. (1992) System condition monitoring – an overview. *Noise .and Vibration Worldwide*, Elsevier Advanced Tech., Sept 1992, 25–29.

Application of system condition monitoring

7

7.1 INTRODUCTION

We have considered the general methods of monitoring hard and soft changes in Condition Monitoring and related diagnostics. As a case study showing the application of these ideas to a practical problem the condition monitoring of a machine tool is considered in detail.

The test vehicle is a flexible machining centre (FMC) Type V4–6 manufactured by Wadkin Machine Tools Ltd. The general layout is shown in Fig. 7.1, while the major sub-assemblies are listed in Fig. 7.2. To determine which sub-assemblies to choose to illustrate the methods, data supplied by the National Council for System Reliability (NCSR) was examined. From their ongoing survey of machine tool breakdowns, see Fig. 7.3, two subsystems have been selected to illustrate the problems and possible solutions. They are the coolant system and the 'X' axis drive. The coolant system is a useful system to choose to illustrate hard fault diagnostics, for its performance is basically described in a binary sense; coolant present or not. The X-axis drive system is a dynamic system which can be used to illustrate the systems approach to condition monitoring. In addition both systems will show that any theoretically based technique has to be used with care. The application of any method has to be adjusted to meet the constraints that the practical situation dictates.

7.2 CM OF HARD FAULTS

Hard fault diagnostics is concerned with system elements that have, or can be considered to have, two operational states, 'healthy' or 'sick'. It is binary in nature. This chapter examines suitable techniques for such hard

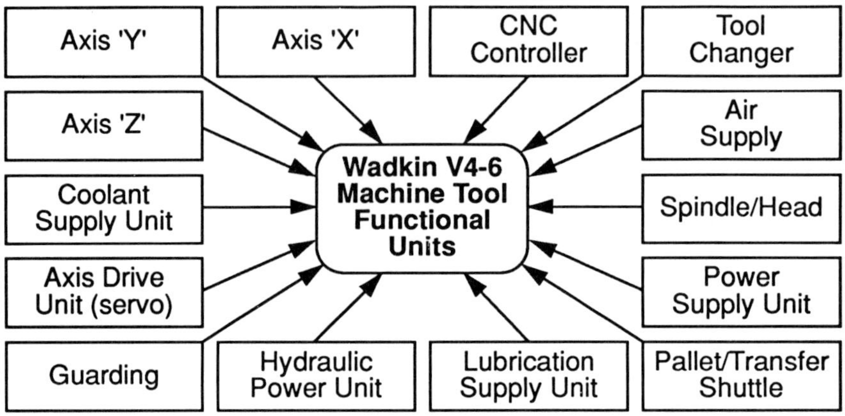

Fig. 7.1 General layout.

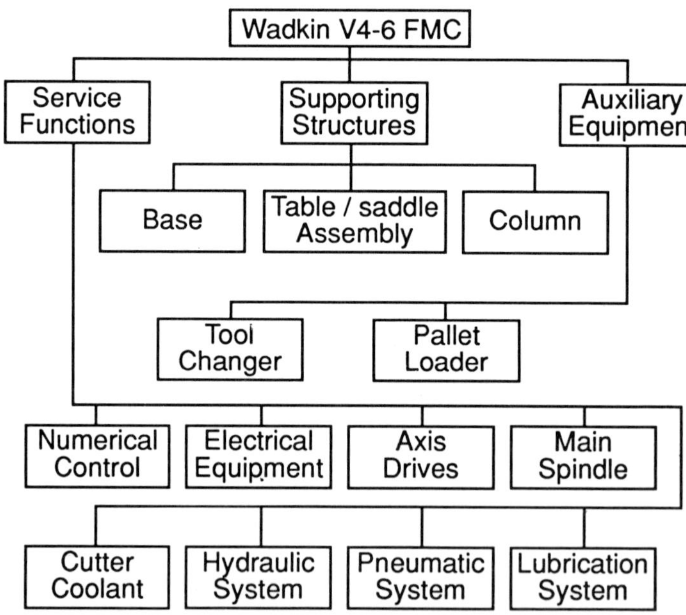

Fig. 7.2 Main sub-assemblies of the machine tool.

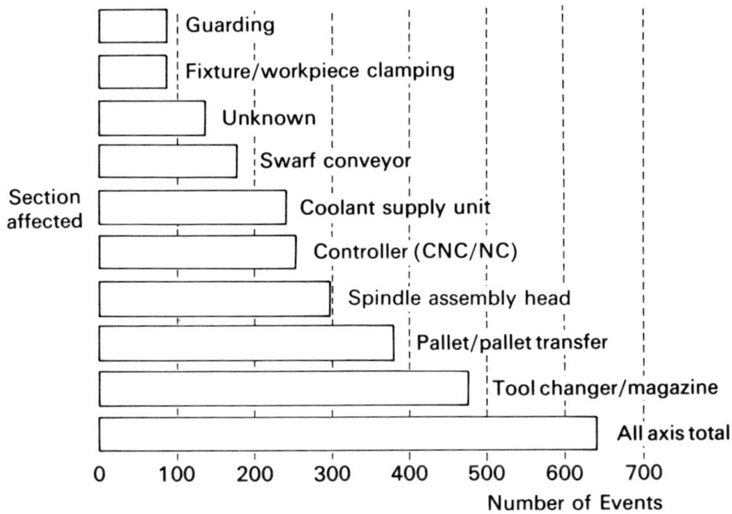

Fig. 7.3 CNC failure modes after NCSR.

faults based on failure mode and effect analysis (FMEA) and fault tree analysis (FTA). A practical application is presented where the method as applied to a coolant system of a machine tool. A coolant system is not complex, is well understood and is an essential subsystem of the machine tool and is thus an ideal vehicle for condition monitoring.

7.3 FAILURE MODE AND EFFECT ANALYSIS

FMEA (Annon, 1990) was developed by reliability engineers for solving problems arising from the malfunctioning of hardware components. Basically, it answers the question 'what if?'. It is concerned in the manner of failure, the effect of such failures, the costs in time and money of repair and other information related to reliability. The functional description of the system and every component has to be derived. This approach is useful for a pattern recognition type of diagnostic system where a set of states is continuously monitored. Once the failure has been identified it can be noted together with the pattern of the monitored states for future reference.

7.4 FAULT TREE ANALYSIS

This method evolved from system safety analysis in the areospace industry in the early 1960s (Lambert, 1977). It traces the individual or sequence of events a particular failure produces. A fault tree is a Boolean logic model based on an FMEA that depicts the parallel and sequential combinations of events that cause a 'top' level fault to occur. Once established the fault tree can be used to determine what sensors are necessary to monitor the top level event or the combination of events. It can also be used via access points and fault matrices as a diagnostic tool which if enough sensor signals were available can be used to locate the fault.

7.5 THE COOLANT SYSTEM

Figure 7.4 shows a schematic diagram of the coolant system (Thorpe, 1989). The system comprises two reservoirs (both 39 litres) located either side of the machine tool base. The motor-driven pump is controlled by means of a switch on the control panel and can be switched off if coolant is not required by the machining operation. Coolant is pumped via solenoid operated valves and a manually operated flow control/shut-off value to the coolant outlets at the spindle nose.

With the coolant pump running, flood and induced coolant is turned off and on by solenoid valves which are energized/de-energized by programmed demands. Coolant flow is manually regulated by means of the control/shut-off tap. Used coolant drains back into the reservoirs where

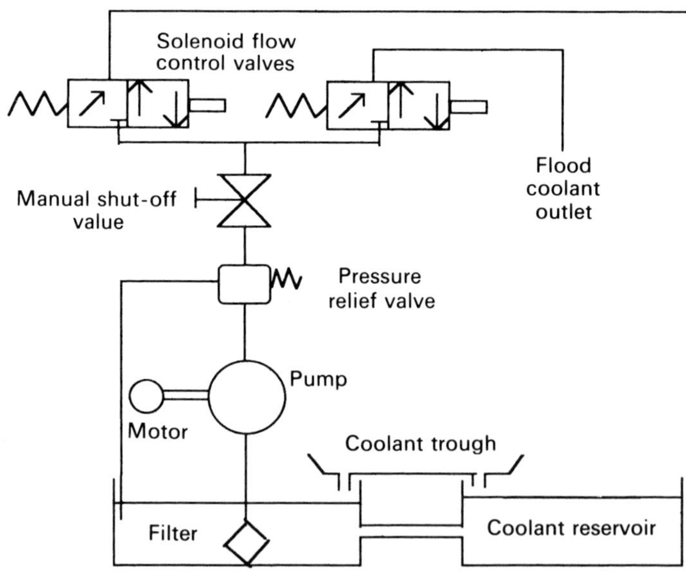

Fig. 7.4 Schematic diagram of the coolant system.

wire-mesh strainers prevent the ingress of large particles of swarf. A replaceable wire-gauze filter is fitted to the pump inlet.

The electrical system operates the pump motor and the two solenoid operated flow control valves. If consists of three relays activated by commands from the controller. Figure 7.5 shows the arrangement.

The functional description of the system is shown in Fig. 7.6. This details the events which occur on receipt of a command in order to achieve the required objective. Inputs C_1, C_2 and C_3 are the energizing/de-energizing demands from the controller for motor, flood coolant and induce coolant respectively.

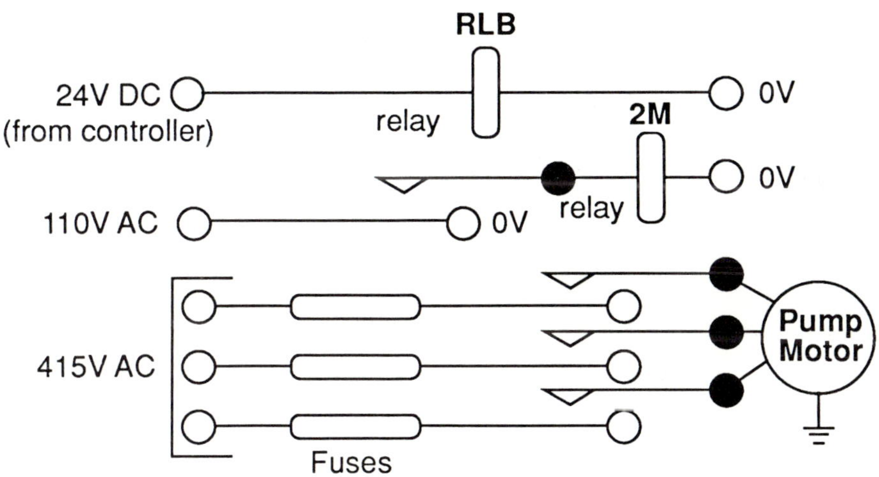

(a) Coolant Pump Electrical System

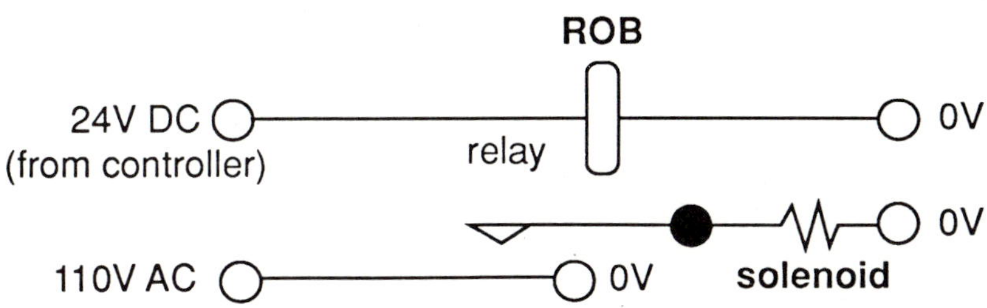

(b) Flood Coolant Electrical System

Fig. 7.5 Electrical schematic of the coolant system.

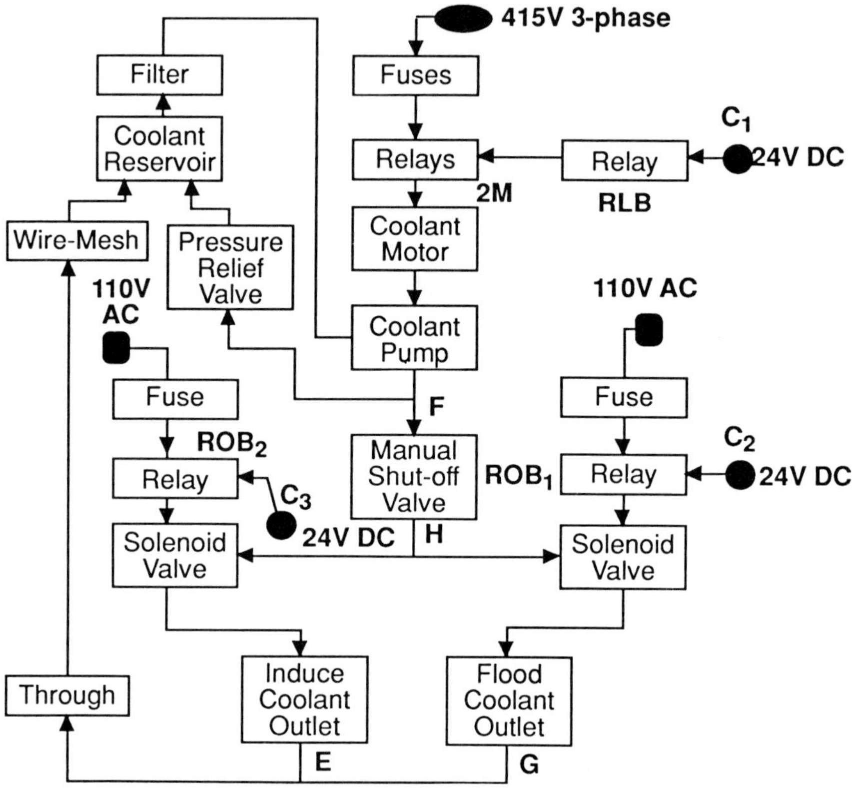

Fig. 7.6 Functional schematic diagram of the coolant system.

7.6 THE METHOD

A system functional check-out of the coolant system was carried out in two stages, when the system was idle and when working. The results are tabulated in Table 7.1. This check-out aimed to identify potential fault areas. The table outlines possible causes of failure in both machine operating modes. This analysis is a prerequisite for fault tree creation.

An FMEA based on the functional description diagram was carried out and the results are listed in Table 7.2.

The FMEA was used to establish the flood coolant supply fault tree and is shown in Figs. 7.7 and 7.8. The fault tree for the induced coolant supply can be regarded as similar and this is not considered further. Two of the failure modes described in the fault tree, namely misdirected coolant and

Table 7.1 Coolant system fault classifications

Cutter coolant activity	*Possible fault areas*
Waiting for command (idling)	Low coolant supply Poor coolant quality Coolant jet misdirected Coolant pump-set fail to start Coolant pump-set operating condition u/s
Supplying coolant (operating)	Pump-set failure Low coolant pressure Low coolant supply Low coolant flow-rate Relay failure Valve failure Coolant jet misdirected due to collision Power failure

coolant quality, are not directly attributable to a single cause. Both faults are outside the field of hard faults.

Most modern systems consist of various modules that are replaced as complete units. These are commonly referred to as Line Replaceable Units (LRUs). Hence it is apposite to identify the unit at fault rather than the specific malfunction within the fault. Provided the unit can be isolated this is sufficient at 'field' level where the unit is replaced and repair can be carried out off-line. For example, no attempt was made to identify faults within the pump-set as the coolant pump and motor are replaced as a complete unit. This approach may be of little benefit to repair personnel but is adequate and efficient in a condition monitoring environment.

The identified LRUs of the coolant system are:

- relays
- coolant pump-set
- solenoid valves
- coolant reservoir
- pressure relief valve
- manual shut-off valve
- filter.

An access point matrix, shown in Table 7.3, was established which lists the accessible signals available for monitoring and fault diagnosis.

The matrix describes the relationships between the system inputs and outputs. For example, energizing coolant pump-set (system input) would set voltage at access point at coolant pump-set, relay RLB and relay 2M

Table 7.2 Failure mode and effect analysis of the coolant system

Items	*Function*	*Possible hard fault*	*Possible failure mechanism*	*Final effect*	*Possible ways of detection*	*Suggested monitor sensor*
Manual control valve	Regulate pressurized coolant to the solenoid valves	1. Fail to open 2. Fail to shut	1. Damaged valve 2. Operator's fault	No coolant to cutter and workpiece	1. Coolant pressure in the system 2. Coolant flow at outlet	1. Pressure sensor 2. Flow detector
Solenoid control valves	Control flow of coolant to outlets (flood and induced)	1. Fail to open 2. Fail to shut	1. Damaged valve 2. Faulty electrical connection	No coolant to cutter and workpiece	1. Coolant pressure in the system 2. Coolant flow at outlet	1. Pressure sensor 2. Flow detector
Pressure relief valve	Safety device	1. Fail to shut when maximum flow needed 2. Fail to open when max. press. reached	1. Damaged valve 2. Incorrect spring setting	1. Possible insufficient pressure at outlet 2. High pressure in piping leaks	Coolant pressure in the system	1. Pressure sensor 2. Flow detector
Coolant pump set	Supply pressurized coolant	(a1) Pump motor fails to switch-on (a2) Pump fails to rotate	(a1) Damaged pump set (a2) Faulty electrical components	(a) No coolant to cutter and workpiece	Coolant pressure in the system	Pressure sensor

		(a3) Pump set fails during operation (b1) Pump fails to provide adequate pressure	(a3) Broken bearing/shaft (b1) Various	(b) Insufficient coolant pressure		
Coolant filter	Prevent ingress of particles into coolant pump	1. Filter blocked 2. Allow particles through	1. Damaged filter 2. Too many particles or swarf in coolant reservoir	1. Loss of coolant pressure/flow 2. Damage coolant pump set	1. Coolant pressure in the system 2. Coolant flow 3. Vibration from the pump set	1. Pressure sensor at pump outlet 2. Flow detector 3. Accelerometer
Wire-mesh	Prevent ingress of large particles and swarf into the coolant reservoir	1. Blocked 2. Allow particles and swarf through	1. Damaged valve 2. Swarf not periodically removed	1. No coolant return to reservoir 2. Swarf and particles build-up in the reservoir	Coolant pressure in the system	1. Pressure sensor 2. Flow detector
Relays	Close or open switch	Faulty contacts	1. Faulty electrical components 2. Mechanical contact failure 3. Electrical power failure	Loss of coolant supply	1. Electrical signal (high and low) 2. Pressure signal 3. Flow detector	1. Electrical signal monitoring 2. Pressure sensor 3. Flow detector

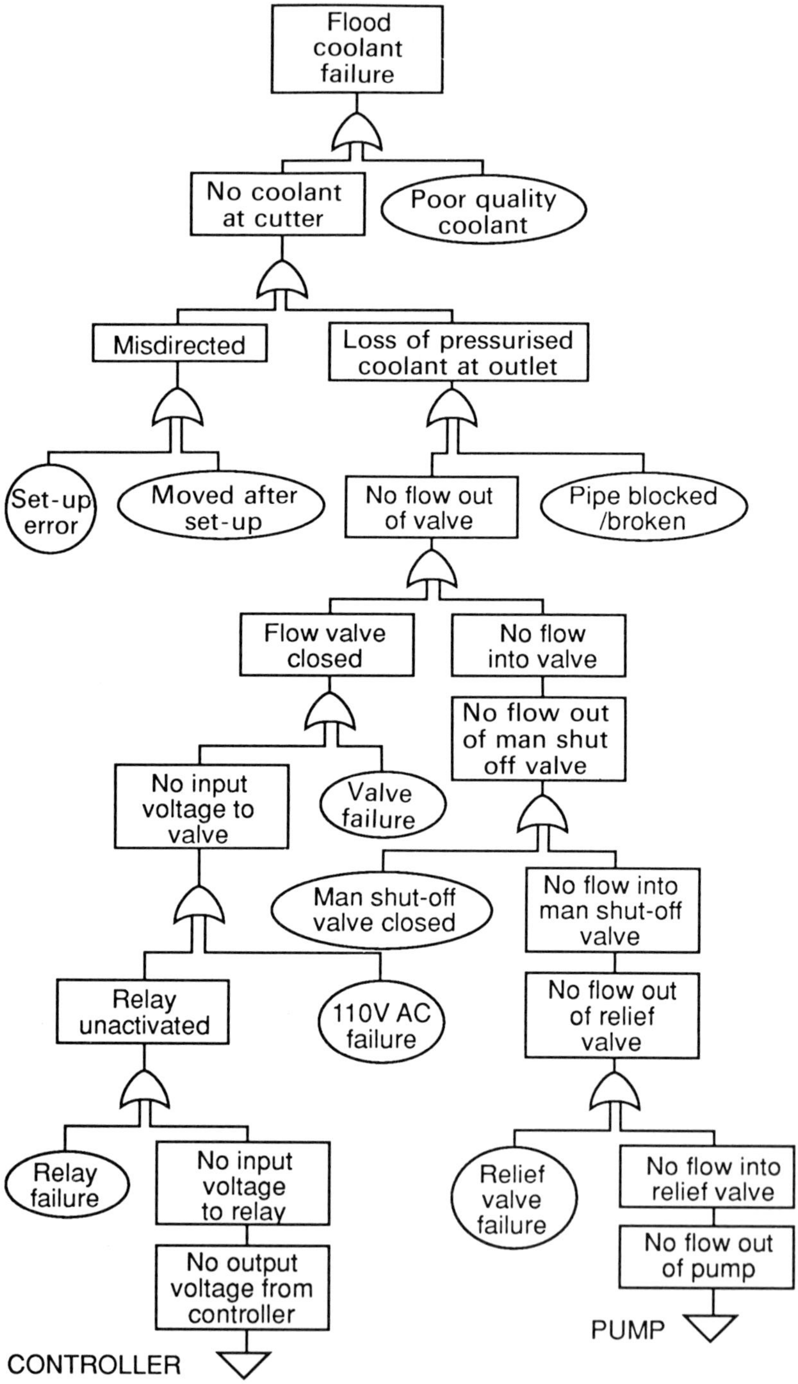

Fig. 7.7 Flood coolant supply fault tree.

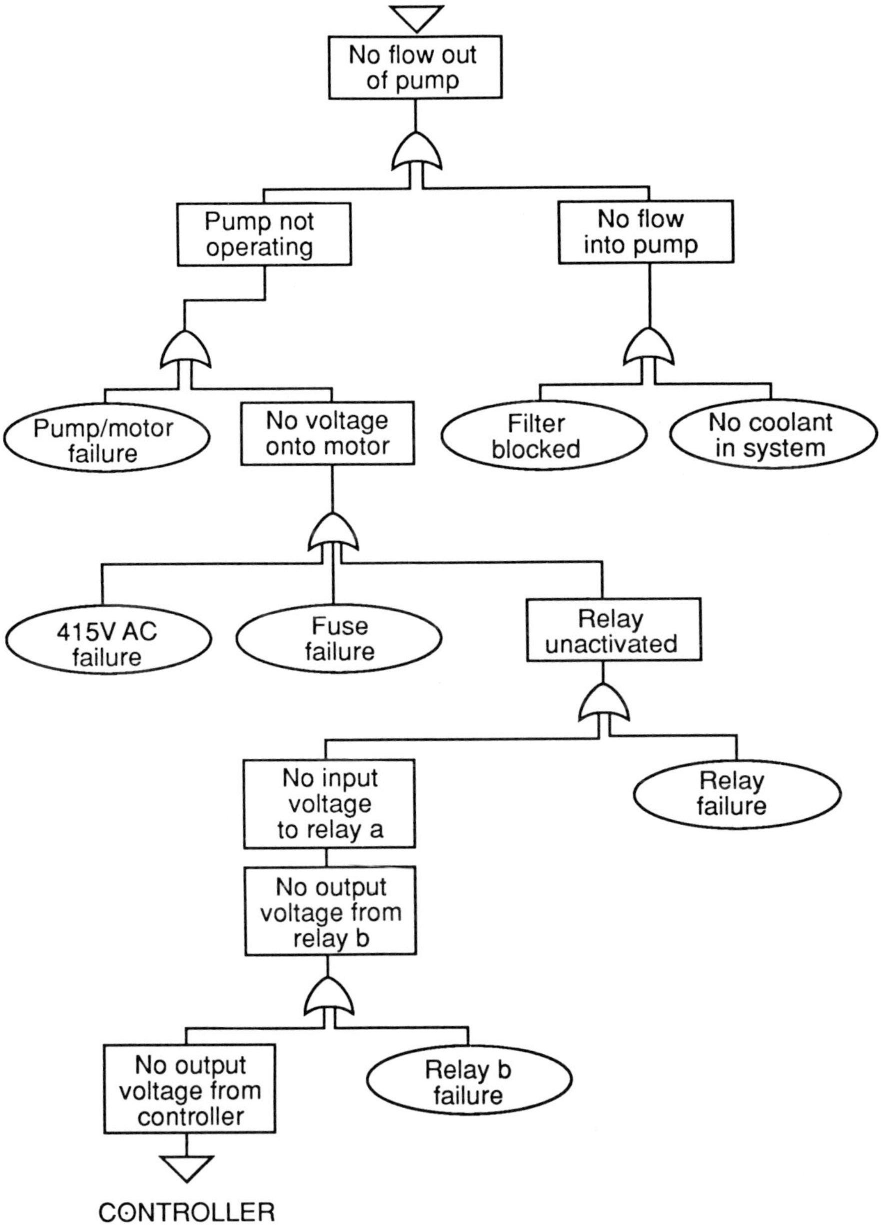

Fig. 7.8 Pump-set fault tree.

Table 7.3 Access point matrix

Event → *↓ Access point*	*Energize coolant pump-set*	*Energize flood coolant*	*Energize inducer coolant*
Coolant pump	1	1	1
Flood solenoid	0	1	0
Inducer solenoid	0	0	1
2M relay	1	1	1
RLB relay	1	1	1
ROB1 relay	0	1	0
ROB2 relay	0	0	1

high and all others low. While there are a number of existing access points for electrical faults there are none for mechanical faults. It was therefore necessary to add access points and sensors to aid mechanical fault detection. Three additional parameters were included: coolant level at the reservoir, outlet pressure of the coolant pump-set and flow detection at each coolant outlet. These access points are marked E,F,G and H on Fig. 7.6.

An updated access point matrix is shown in Table 7.4. It represents the condition when the motor and coolant control valves are energized simultaneously. The matrix shows the expected pattern of '1's and '0's under fault-free operation and indicates by 1 or 0 values to ensure a healthy system. A '1' in the matrix indicates that a relay output is activated, pressure is up, flow is taking place and the reservoir level is correct. From this access point matrix a fault matrix was established and is shown in Table 7.5.

Table 7.4 Updated access point matrix

Event → *↓ Access point*	*Energize coolant pump-set*	*Energize flood coolant*	*Energize inducer coolant*
Coolant pump	1	1	1
Flood solenoid	0	1	0
Inducer solenoid	0	0	1
2M relay	1	1	1
RLB relay	1	1	1
ROB1 relay	0	1	0
ROB2 relay	0	0	1
*Coolant level	1	0	1
*Pressure at F	1	1	1
*Pressure at H	1	1	1
*Flow at G	0	1	0
*Flow at E	0	0	1

*Indicates that additional sensors are needed

Table 7.5 Updated fault matrix

Row ID letter	*Coolant level*	*Relay RLB*	*Relay 2M*	*Relay ROB*	*Pressure at F*	*Pressure at H*	*Flow at G*	*Most prob. failure*
A	1	0	0	1	0	0	0	Relay RLB fail to close
B	1	1	0	1	0	0	0	Relay 2M fail to close
C	1	1	1	0	1	1	0	Relay ROB fail to close
D	1	1	1	1	0	0	0	Motor/pump set failure
E	1	1	1	1	1	1	0	Flood solenoid failure
F	1	1	1	1	1	0	0	Man. control valve shut
G	0	1	1	1	0	0	0	Liquid level low
H	0	1	1	1	1	1	1	Coolant level sensor failure
I	1	0	1	1	1	1	1	Relay RLB sensor failure
J	1	1	0	1	1	1	1	Relay 2M sensor failure
K	1	1	1	0	1	1	1	Relay ROB sensor failure
L	1	1	1	1	0	1	1	Pressure sensor at F failure
M	1	1	1	1	1	0	1	Pressure sensor at H failure

7.7 FAULT LOCATION

The fault tree diagram highlights the critical signals to be monitored. It also identifies the redundant signals and so focuses monitoring only on the critical signals. The Fault Tree Analysis has revealed that there is a single overriding fault occurrence that can be used to monitor the health of the coolant system.

By continuously monitoring the output flow sensor at 'G', see Fig. 7.6, during times when the coolant is required as requested by the controller, reliable unmanned operation is possible, as far as the coolant is concerned. It must be emphasized that not all possible failure modes are represented in Table 7.5. To do that the fault matrix would have to contain all the

Table 7.6 Fault matrix

	1	*2*	*3*	*4*	*5*	*6*	*7*
A	1	0	0	1	0	0	0
B	1	1	0	1	0	0	0
C	1	1	1	0	1	1	0
D	1	1	1	1	0	0	0
E	1	1	1	1	1	1	0
F	1	1	1	1	1	0	0
G	0	1	1	1	0	0	0
H	0	1	1	1	1	1	1
I	1	0	1	1	1	1	1
J	1	1	0	1	1	1	1
K	1	1	1	0	1	1	1
L	1	1	1	1	0	1	1
M	1	1	1	1	1	0	1

combinations of the ‘n’ entries, where n is the number of columns in the matrix. Only the recognizable LRU failures have been included.

The FMEA and the FTA are used to produce the fault matrix. It is the fault matrix which gives the failure signatures for, in this case, 13 identifiable single fault conditions. The matrix holds the information that enables the efficiency of diagnosis to be examined. The heart of the matrix is the bit pattern, which is shown in Table 7.6 in shortened form.

Each of the 13 faults listed has a unique signature or bit pattern. So if any fault occurred singly then there is a high confidence that the faulty LRU would be identified. If more than one fault occurred at the same time then it could be possible to identify the faults. However, for success the resulting bit pattern must be different from any other entry and cannot be duplicated by any other combination of two faults. For example, if both the 2M and RLB relays were faulty one would have the combined signatures of 1,0,0,1,0,0,0 (row A) and 1,1,0,1,0,0,0 (row B) to give 1,0,0,1,0,0,0 and this pattern is not repeated in the matrix. However row A has an ‘obliterating’ effect. Row A combined with any row from B to J produces the same pattern, 1,0,0,1,0,0,0. If one proceeds through the matrix, taking two at a time combinations, one sees that multiple faults cannot sensibly be identified. Multiple faults can occur independently which is highly unlikely or in cascade, i.e. a ‘knock-on’ effect. In general care should be taken when analysing hard fault diagnostics to ensure that the system does not have the capability of suffering ‘knock-on’ effects where a single failure by its very nature causes other faults to occur. However, in practice for a system such as the coolant system single faults are much more likely than multiple faults.

7.8 THE X-AXIS DRIVE SYSTEM

From the NCSR review the axis systems suffer the greatest number of breakdowns. This was true of the system under consideration, the Wadkin V4–6 FMC. Figure 7.9 presents a schematic of the x-axis drive system. Fundamentally it's a position servo moving at velocities determined by the controller. The motor drives a screw through a belt and pulley arrangement. The ballnut, threaded onto the screw, is attached to the bed of the machine thus achieving linear translation. Figure 7.9 includes the major sub-assemblies of the electronic servo-control.

This x-axis drive is a dynamic system, and as such, system-based principles were used to design and implement a condition monitoring scheme. The general approach was then modified so as to produce a practical solution.

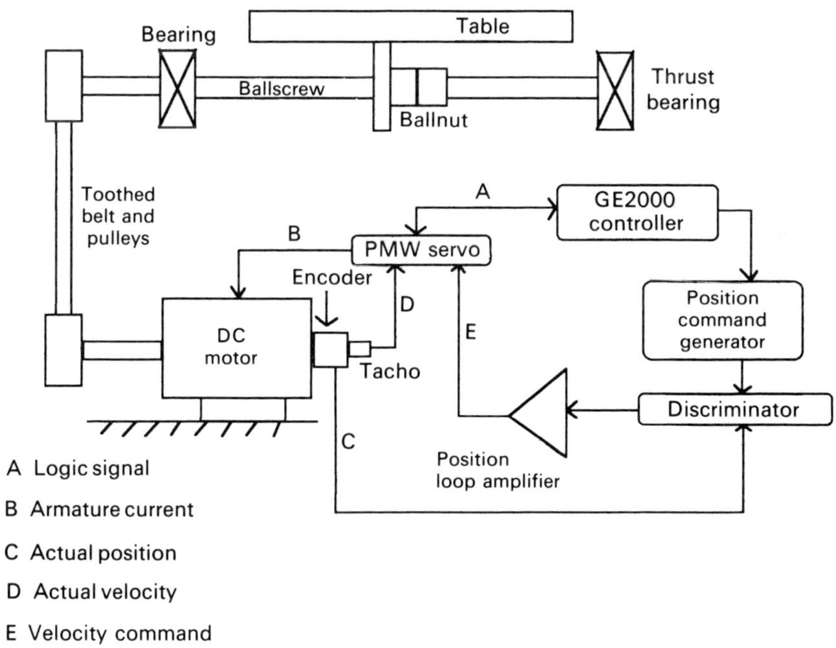

Fig. 7.9 X-axis drive schematic.

7.9 SYSTEM ACCESS POINTS

The motor drive control subsystem provided a convenient source of signals that could be classed as 'input' and 'output'. These were the velocity command and the motor's armature current. The velocity command signal was tapped-off at the servo control unit, passed through a passive low-pass filter (15 Hz) and then stored on a digital storage oscilloscope (DSO). The cut-off frequency, 15 Hz, was chosen to match the possible maximum feedrate of the axis drive motor which is 150 revs/min for the ballscrew (or 62.5 for the tacho ripples). Details of the data filter are given in Fig. 7.10. The armature current was tapped-off from the built-in current transducer. This signal is passed through a filter similar to that used on the input and then to the DSO. The use of similar filters effectively removes them from any consideration in the modelling. Basically they 'cancel' each other out. Both signals were easily accessible and needed no additional sensors. The filtered signals showed ripple characteristics of the servo system. These are:

- Motor torque ripple = 109 cycles/rev
- Tachometer ripple = 25 cycles/rev
- PWM frequency = 2000 Hz (approx) (PWM is pulse width modulation)
- Triangle wave generator = 1000 Hz (approx).

The ripples were smoothed out using a least squared technique (Savitzky *et al.*, 1964).

The DSO can be programmed to control the signal amplification. The sampling interval was set at 1msec. The data stored was transferred directly via an IEEE interface to an IBM AT type computer which acted as the supervisory computer for the total data acquisition and analysis process. The schematic of the experimental set-up is shown in Fig. 7.11.

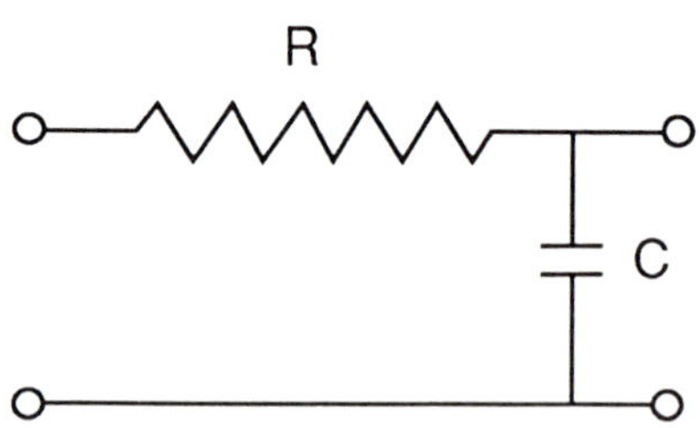

Using CR = 1/2 πf where C = capacitor in μF, R = resistor in Ω.
Fixing C = 1 μF, f = 15Hz, then R = 10^6/[(2(15) (1)] = 10.6 kΩ.
Using bandwidth x rise time = 0.35 secs, rise time = 0.35/15 = 0.0233.

Fig. 7.10 Passive low pass filter design.

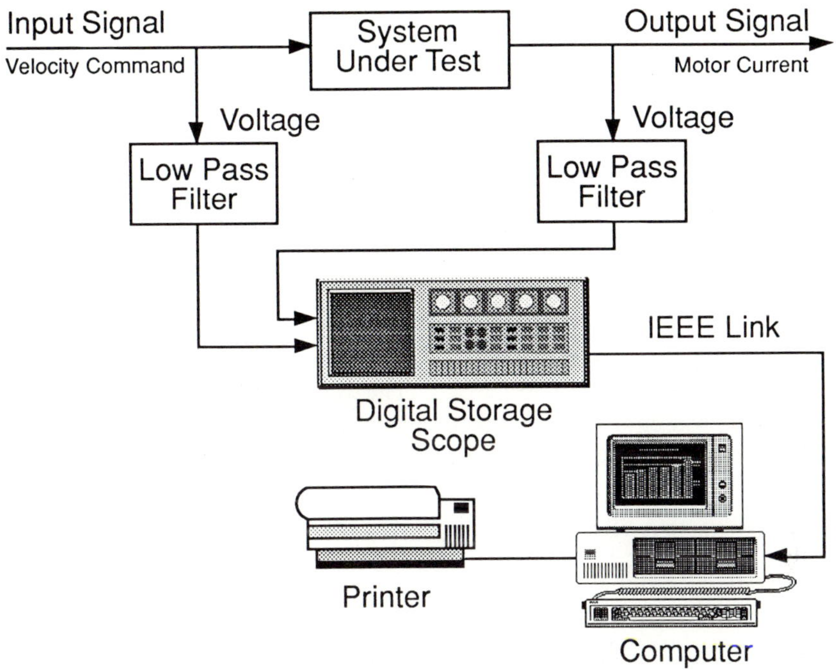

Fig. 7.11 Schematic of experimental set-up.

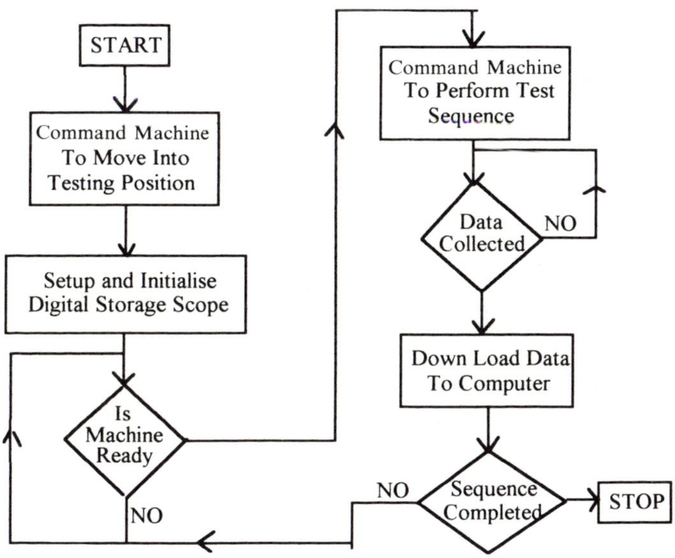

Fig. 7.12 Flow chart of experimental data gathering.

Data was automatically collected over long periods of unmanned operation. The machine controller (GE2000) communicated to the supervisory computer via a serial link whenever it encountered a customized 'M' code (Wadkin, 1983) placed in machine part programs. The supervisory computer then used this signal to synchronize the action of the machine and the data logging and data processing. The software flow chart is shown in Fig.7.12.

7.10 DYNAMIC MODEL

The velocity command signal rises to a voltage level which is a linear function of the velocity requested. The demand can be conveniently modelled by a first order system response to a step input. The motor current is a step-like response consisting of a large initial rise and fall followed by an oscillatory steady-state response, see Fig. 7.13. A series of experiments were undertaken to identify a suitable dynamic model of the system that was considered to be the transfer between the velocity command as input and the motor current as output. Responses were measured over various times and at various velocity demands. The input and output data were recorded and the MATRIX_x computer package (Integrated Systems Inc, 1986) used to estimate linear time-invariant transfer function models. These models are listed in Table 7.7. (The package actually uses a least squares method to fit a z-domain transfer function

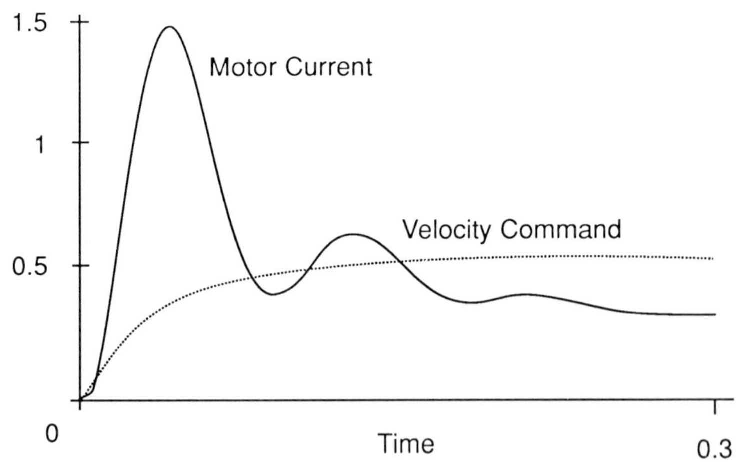

Fig. 7.13 Typical input and output signals.

Table 7.7 Transfer functions fitted to velocity demand (input) and motor current (output)

Speed (mm/min)	*Model fitting span(secs)*	*Transfer function*
300	0.06	$\frac{0.59s^2 + 1801s + 34477}{s^2 + 51.1s + 12639}$
300	0.30	$\frac{0.55s^2 + 1030s + 10604}{s^2 + 35.95s + 6950}$
500	0.06	$\frac{0.87s^2 + 2283s + 17760}{s^2 + 79.99s + 16851}$
500	0.30	$\frac{0.68s^2 + 1368s + 8983}{s^2 + 52.61 + 10852}$
700	0.06	$\frac{1.91s^2 + 2382s + 13654}{s^2 + 90.24s + 19123}$
700	0.30	$\frac{0.93s^2 + 1386s + 8136}{s^2 + 53.24s + 12189}$
1000	0.06	$\frac{1.46s^2 + 3899s + 13354}{s^2 + 136.3s + 30156}$
1000	0.30	$\frac{0.97s^2 + 2233s + 10303}{s^2 + 68.48s + 19088}$

which essentially is an auto-regressive moving average time-series model (Box and Jenkins, 1970). Having obtained the z-domain model its s-domain equivalent can be estimated by using a bilinear transformation.) The models were fitted over two time spans. This illustrates the compromise in modelling a dynamic system in incorporating the two basic response regions; the initial transient part and the steady-state part.

Models of fourth, third and second order were investigated. Little appeared to be gained by going beyond the second order. In other words, the transfer function is a ratio of two quadratic polynomials. (While the numerators of the models were quadratics they all were dominated by a first order factor. This 'differentiating' effect can be observed by the initial large rise in the response.) This model structure was found to be the most appropriate for the given input/output signals.

While the models are useful in simulating the response of the x-axis drive assembly for diverse inputs they are of little use, in this instance, as a basis for a condition monitoring system. The difficulty is that the coefficients of these 'good' models do not relate to any particular feature or parameter of the physical system. Therefore only the response signal was used as a fingerprint of the behaviour of the drive.

7.11 OBSERVABLE FAULTS

Temperature variation is well known as a cause affecting machine tool response changes. In order to eliminate the temperature effect it is necessary to test when thermal-equilibrium had been reached. It was not possible to control temperature but in order to produce near thermal-equilibrium the machine tool was set working over a pre-arranged sequence for at least an hour before any tests were carried out.

Other conditons are necessary for valid results. In particular during the time between recording a reference level and a test for faults no significant changes should occur which might affect the steady-state level of the motor current. To this end tests were carried out with no fixture on the machine table. This condition is not too restrictive for it corresponds to the time during pallet changes.

Before the output response of the drive could be used as a condition monitoring measure the nominal response of the drive had to be establish-

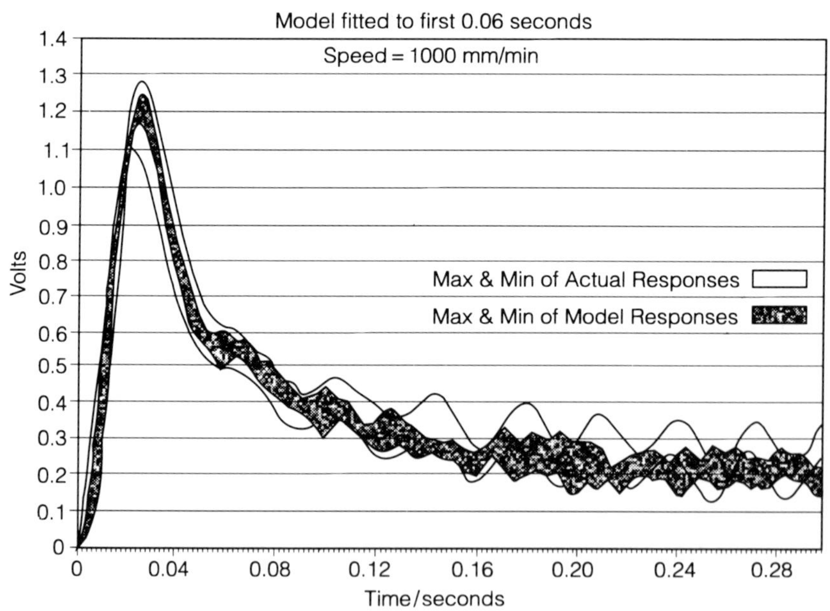

Fig. 7.14 Model output using nominal tolerance band.

ed. This was done by repeatedly traversing the table in both directions, x+ and x−. It emerged that the response was dependent on direction. While the differences were small, good practice dictates that the signal should only be monitored when the table moved in one particular direction; this was what is defined as x+. Thirty runs were executed in the x+ direction and the envelope of the measured responses established. This nominal tolerance band was used as the reference for system checkout and is shown in Fig. 7.14 together with the model's output.

Because it was not possible to await the development of fault conditons in the machine tool, faults had to be simulated. The faults considered were:

1. varying amplifier gain: it simulates the effect of amplifier drift;
2. varying ballnut preload: it simulated the effects of (a) ingress of dirt; (b) increase of friction in ballscrew or slides; (c) onset of backlash and (d) wear and tear;
3. varying system backlash: this was done in two ways:
   (a) reducing end-thrust bearing preload to cause backlash: this also simulates slight loss in axis system preload;
   (b) reducing the tension of the transmission belt to cause backlash; this simulates backlash due to belt slackness.

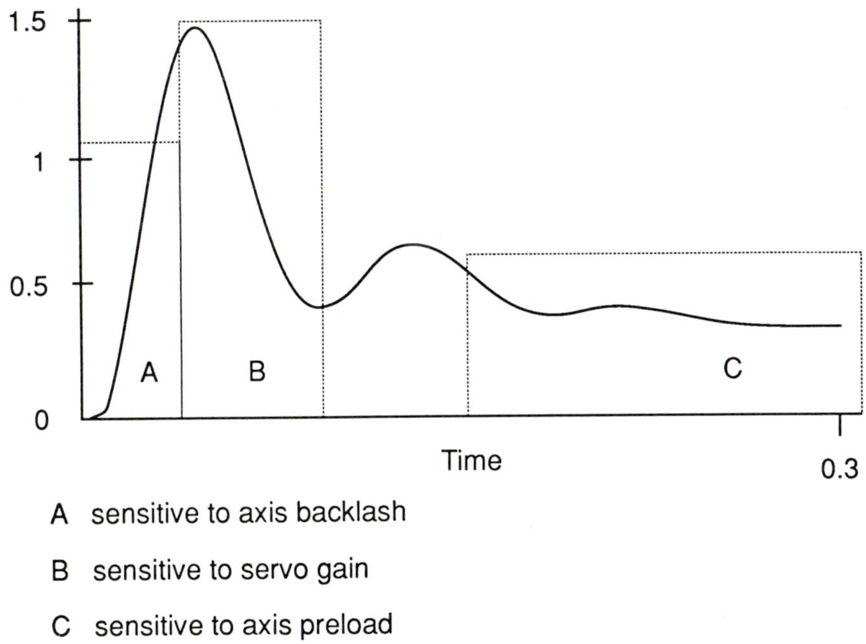

Fig. 7.15 Fault regions of sensitivity.

Transient: Capture motor current transient response
Smooth the data using a 15 Hz passive low pass filter
Sample rate = 40 k/sec

Steady-state: Capture motor current steady state response
Sample rate = 80/sec
Data processed to give result at 10 mm intervals
(one per rev. of ballscrew).

These basic drift changes were found to be observable from the motor current. It was found that these 'faults' manifested themselves in three areas of the motor current response, see Fig. 7.15.

The test schedule that was used to monitor these areas is shown in Table 7.8. The machine table positions are given in units of total travel.

7.11.1 Amplifier gain

The dynamic response performance of the motor current was used to monitor the health of the servo unit. The response at 3500 mm/min feedrate was used. The steady-state response was not required. The axis servo is a 'live servo' system which exhibits a 'stationary current' even when the axis table is not moving. This characteristic influenced the way the current data was collected. The obvious approach was to set a threshold value for triggering the data collection procedure. The threshold was set at 0.7 volts.

There are two adjustable resistors in the servo unit for users to select the appropriate Gain and Balance (bias) to suit the system. These two

Table 7.8 Test procedure for axis drive system

Fault monitored	*Test procedure*	*Signal analysis*	
		Transient	*Steady-state*
Drift to servo gain	1. move to x = 250, stop 2. move to x = 300, stop 3. trigger data collection move to x = 350 feedrate = 3500 mm/min	Yes	No
Increase in backlash	1. move to x = 350, stop 2. move to x = 250, stop 3. trigger data collection move to x = 350 feedrate = 500 mm/min	Yes	No
Drift in preload	1. move to x = 0, stop 2. trigger data collection move to x = 550 feedrate = 800 mm/min	No	Yes

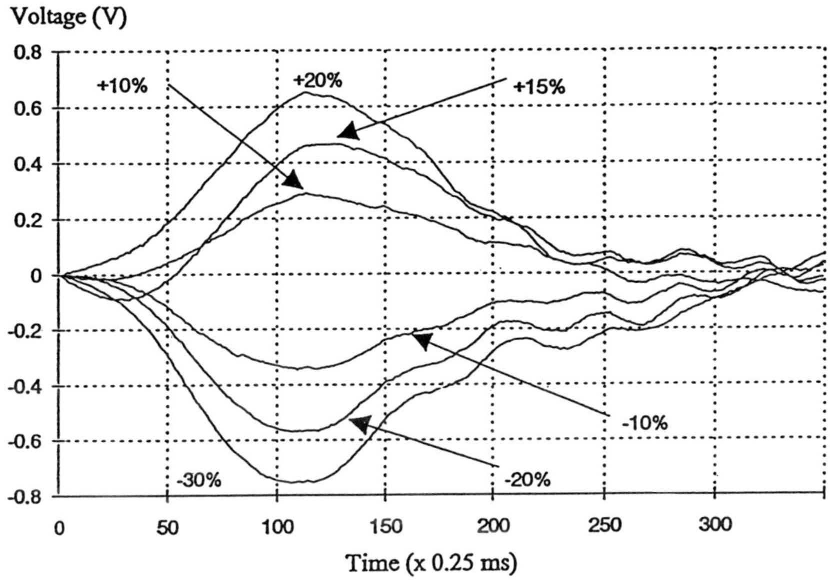

Fig. 7.16 Servo gain test – mean deviation vectors.

resistors are normally set by the manufacturer. For each setting 30 runs were carried out with the transient signal recorded and averaged. The spread about the average for these signals is small ($<5\%$) and so it is clear that drift changes in gain are easily observed. The gain was changed from its nominal over the range -30% to $+20\%$. Typical envelope and deviation responses are shown in Fig. 7.16.

The full range of bias adjustment is only 8%. It was found that the discernible changes in response due to bias drift were so small as to prove an impractical parameter to monitor.

7.11.2 Ballnut preload

The rotational motion of the motor is transmitted via pulleys and timing belt to the ballscrew which is then converted to linear motion. The table, which is fitted with a linear bearing, is then driven along the axis sliding on the axis slideways. Four strategically placed limit switches protect the machine from over-travel during start-up as well as providing position information to the controller for reference purposes.

The ballscrew is of the recirculating ball type and manufactured to a high accuracy. They are preloaded to give a particular stiffness and backlash free operation. The ballscrew assembly is secured to the main structure of the machine by a preloaded end-thrust bearing assembly. The preload of the ballscrew assembly is achieved by adjusting the annulus vernier gear thereby producing an angular displacement of the nuts about the axis. The preload was changed by disassembling the drive system and rotating the differential gear of the ballnut. The amount of rotation was noted as the number of 'teeth' intervals rotated. Negative (−) is taken to mean a decrease in the preload whilst positive (+) is taken as an increase. Each tooth movement produced a change of 1.5 μm in the tightness of the axis.

A study was carried out to establish the preload along the axis. A strain-gauge bridge placed on the adjustment nuts enabled the profile or fingerprint of the screw to be determined. The steady-state motor current response at 800 mm/min feedrate was used. The transient response was not required. This profile is shown in Fig. 7.17. This strain-gauge sensor, whilst effective, is an additional component to the total system which could reduce the overall system reliability. Of interest was whether this profile could be observed in the motor current characteristics. This was found to be true thus eliminating the need for an additional sensor to monitor this key parameter. Figure 7.18 shows the linear relationship between the strain-gauge output and the measured motor current.

Experiments were undertaken by successively adjusting the vernier gear

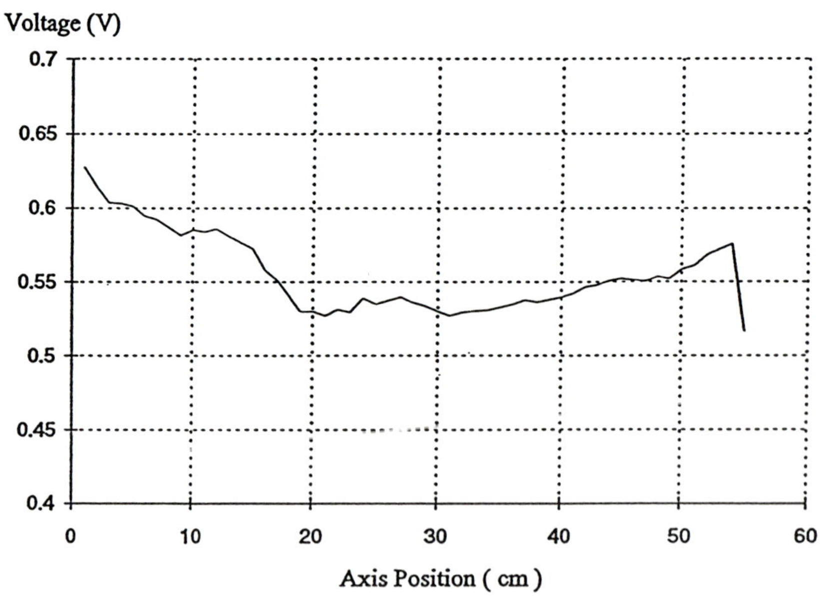

Fig. 7.17 Motor current profile.

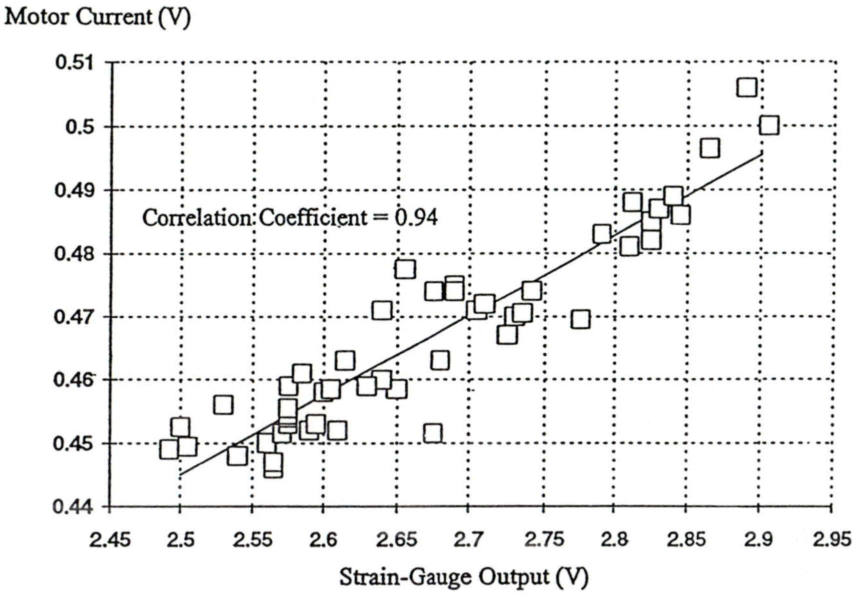

Fig. 7.18 Axis correlation for ballscrew profile.

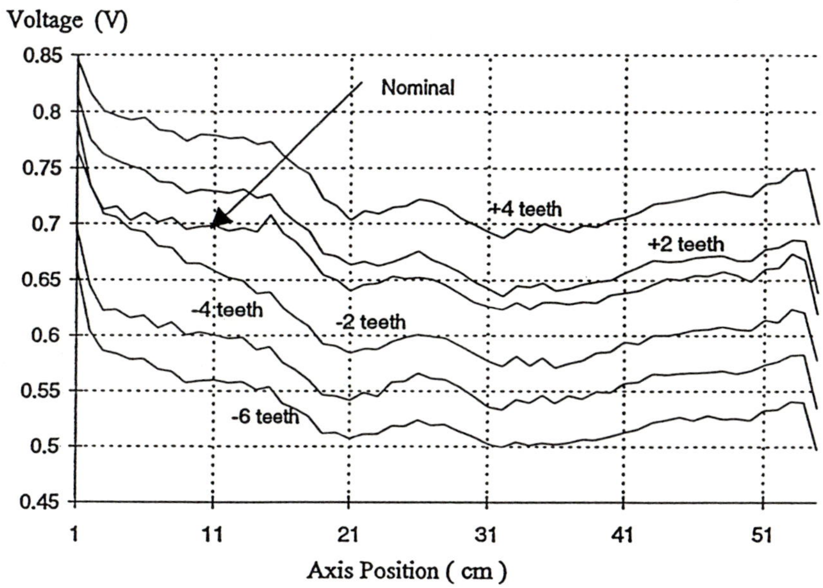

Fig. 7.19 Axis preload test – motor current profiles.

so as to produce various values of preload. These variations simulate faults in the ballscrew due to wear and ingress of foreign matter. Figure 7.19 shows the profiles for various settings indicating that the strain-gauge sensor is indeed not required and that this physical fault can be monitored using the motor current.

7.11.3 System backlash

It was found that the transient portion of the motor current response shows the condition of axis backlash only when performed with a suitable machine test procedure. This standard test (Wadkin, 1983) was used with a feedrate of 500 mm/min.

This axis drive system was disassembled and the ballscrew removed from the machine tool before any adjustment could be made. Backlash was introduced into the axis in two ways: first by the toothed belt and second by the end-thrust bearing. The transmission belt was adjusted by loosening or tightening the pulley and the end-thrust bearing was adjusted by loosening or tightening the end-thrust bearing nut at the end of the ballscrew. To effect these adjustments the axis drive system was modified to change the backlash value of 4 μm to 18 μm. The manufacturer's specification gave 24 μm as satisfactory. This modification involved the

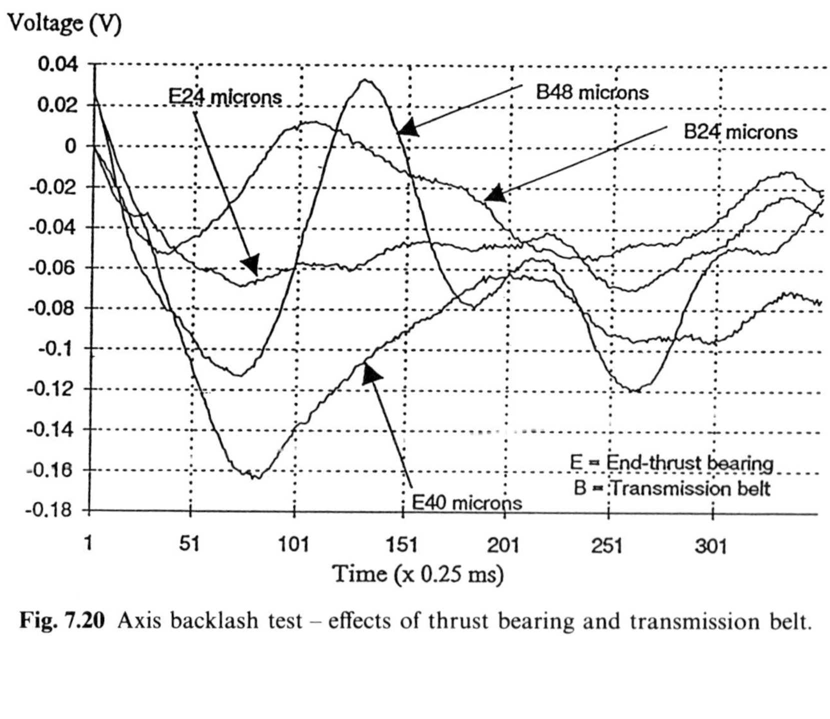

Fig. 7.20 Axis backlash test – effects of thrust bearing and transmission belt.

insertion of two metal strips to allow movement of the actual ballscrew. It was found that the preload had been reduced by 8%.

Starting with a value of 18 μm the belt was adjusted to give backlash values of 24 μm and 48 μm. Subsequently the belt was adjusted back to its original setting and the end-thrust bearing adjusted to give 24 μm and 40 μm. For all settings, including the setting taken as nominal (18 μm), the transient response was measured. Deviation vectors are shown in Fig. 7.20. It is clear from the figure that an 'out-of-spec' backlash condition is observable and in this application faults due to the belt and end-thrust bearing are distinguishable.

7.12 CONCLUSIONS

It has been established that where possible, i.e. the axis drive rather than the coolant system, a system approach using no additional sensors is a practical possibility. In this example the motor current response was sensitive to changes in servo gain, axis preload and axis backlash.

It was a qualified success in that if there is backlash then the monitoring system is capable of monitoring changes in preload and distinguishing between backlash caused by the transmission belt and the end-thrust bearing. However if there was little backlash then detected preload changes could be due to the ballnut or the end-thrust bearing. Either way the principal result is that these mechanical characteristics can be successfully monitored and identified using no additional sensors. Here the system approach relied on only the actual dynamic response of the axis and so has shown that there is little conceptual constraint on using the technique on any dynamic system.

7.13 REFERENCES

Annon (1990) Procedures for performing a failure mode, effect and critical analysis, Department of Defense USA, paper MIL-STD-1629-A, November 1990.

Box, G.E. and Jenkins, G.M. (1970) *Statistical models for forecasting and control*, Holden-Day, San Francisco.

Integrated Systems Inc. (1986) Engineering analysis and control design, *MATRIX$_x$ User's Guide*. Version 5, May 1986.

Lambert, H. E. (1977) Fault tree for diagnosis of system fault conditions, *Nuclear Science and Engineering*, **62**, 20–34.

Savitzky, A. and Golay, M. (1964) Smoothing and differentiation of data by simplified least squares procedures, *Anal. Chem.*, **36**, 8 July.

Thorpe, P. (1989) Condition monitoring of machine tools, *Maintenance*, April 1989, 17–24.

Wadkin (1983) *Wadkin V4-6 Part Programming Manual*, Wadkin Machine Tool Ltd, Leicester, UK.

7.14 BIBLIOGRAPHY

Hoh, S.M. (1992) Condition monitoring of machine tools, Ph.D. Thesis, University of Wales, Cardiff.

Processing diagnostics data | 8

8.1 INTRODUCTION

A condition monitoring system is only of any use if the information gleaned from the monitoring is stored, processed and used by management. Whilst the basic need for a CM system is to furnish the maintenance department with information for repair and sparing it is sensible and good practice to use the information in a planning mode. Other chapters consider the management structure in which CM plays a central role. Here we consider the numerate techniques that can be used to process and present the CM data in a meaningful sense.

Typically, when a fault manifests itself on the shop floor the maintenance operators are called in to effect the repair. After diagnostics, which may be protracted, and repair they usually fill in a form indicating the fault and the nature of the repair. This is where problems start. Often 'convenient' information is given; information that is 'typical' for the machine, such as 'drive problem, bearing replaced'. People don't like filling in forms in any detail and often the information that is given is the minimum that has become 'acceptable'. Further analysis of these records is often limited to gross classifications. Any form of automated CM offers the chance for a much more useful scheme of management reporting. Numerate techniques fall into two fundamental areas; monitoring drifts while the system is still in specification and analysing failure patterns. These are now considered in detail.

8.2 BASICS

The fundamental 'law' of data processing is to process as little as possible to enable an answer to the given problem to be deduced. It must be noted that if raw data is processed in some way, then any decision subsequently made is made on the processed data. For example, a popular technique is to take logs so as to linearize some exponential type law. The resulting fitted straight line is then used as the reference when examining errors. The danger is that characteristics of errors in the transformed data are quite different in the raw data. In addition a typical 'mistake' is having fitted a linear law to the transformed data this law is then re-transformed to the original domain. One must remember that the 'fitting' has **not** been carried out on the raw data. The best advice is to use 'simple' techniques that do little to change the nature of the raw data and are easy to understand. Too often quite sophisticated techniques are used simply because they are available.

The two fundamental approaches in analysing successive values of a particular variable are a histogram or a time-series. Which approach to choose depends on what the analysis is for. The basic difference between the two approaches is that the histogram is static whilst the time-series is dynamic. Consider the example shown in the figure. The two time-

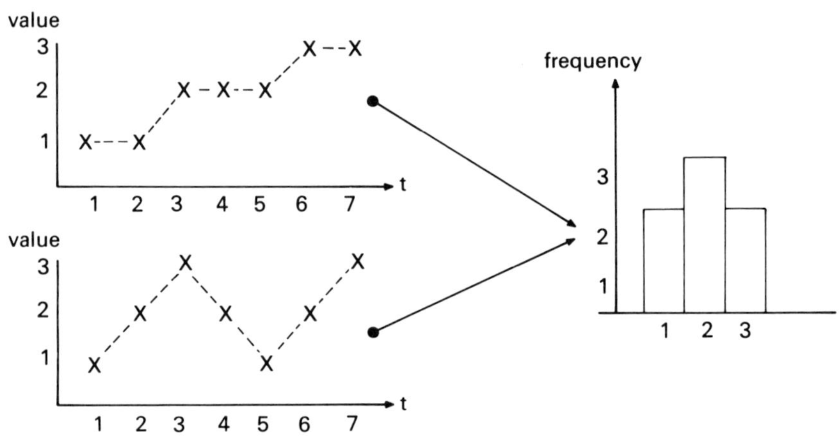

series show quite distinct patterns; one is increasing and the other has a cyclic pattern. Yet the histograms are identical. It must be emphasized that using the histogram is not 'wrong' it is simply that it gives a static snapshot of the process at a particular time and by its very nature does not indicate any dynamic the data may possess.

8.3 STATISTICAL DISTRIBUTIONS

A histogram has two basic measures: its location (mean) and the spread (variance). The mean gives a central value of the variable about which all the data is distributed. The variance is the average squared error where the error is the difference between the data value and the mean. A common assumption is that the data has a Normal or Gaussian distribution (Johanson and Leone, 1964). This makes sense for a number of statistical reasons but this is not the place to consider the theory.

Thus the histogram, whilst discrete, is assumed to be a display of a sample drawn from a continuous distribution. The data in the histogram is used to estimate the mean and variance of the population. The Normal distribution is completely defined by its mean and variance which are the two parameters of the probability distribution function. Whilst the variance is the fundamental measure of spread the practical measure is the square root of the variance which is called the standard deviation. The figure shows the normalized distribution which has a mean of 0 and a standard

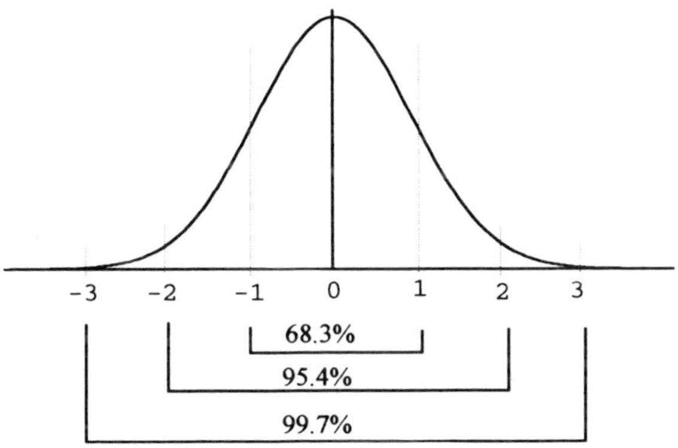

deviation of 1, represented as N(0, 1). Any data that has a Normal, $N(\mu, \sigma)$ distribution can be transformed to this normalized form. This is achieved by subtracting the mean, μ, from every data point and dividing the result by the standard deviation, σ. These new values will then be normally distributed with a mean of 0 and a standard deviation of 1.

Both the histogram and the time-series have a place in CM. The histogram enables a check to be made on the variations that are always present in measurements. Statistical tests are available which enable the estimation of the probability that the sample could have come from the assumed distribution. If that probability is low then one assumes that there has been a change in the system that generated the sample. Crucial to this standard procedure is the establishment of the assumed distribution. This distribution is the reference data which is used to make any decision. It is essential that as much data as possible is collected under a situation where the system concerned is considered healthy and operating under normal operation conditions. In other words the parameters of the distribution are well estimated, which is directly proportional to the amount of data used.

The statistical distribution of data is also used in simulation studies. Here the need is the generation of typical data from a given distribution to produce a time-series. Then each generated set of data will be different but still retaining the statistical characteristics. In other words a typical realization is produced. The figure shows the method. If a probability

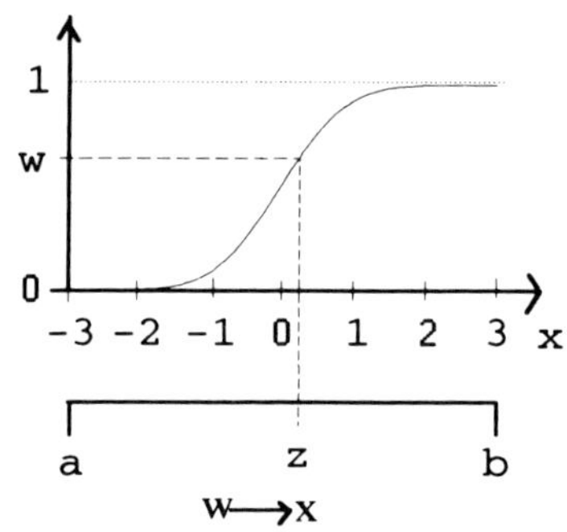

$$z = a + \frac{[x+3][b-a]}{6}$$

distribution is cumulatively summed we get a curve that starts at 0 and is asymptotic to 1. This is because the total probability (the final value of the cumulative sum) must be 1. So if a random number w is drawn from a rectangular distribution (all numbers are equally likely) with a range 0–1, and then w is projected onto the cumulative curve it will generate a number x, say. The example shown in for the normalized Normal distribution, although any distribution can be used. This value of x will have a particular position on the line that ranges from −3 standard deviations to +3 standard deviations. Then the formula shown will produce a value z which represents a number that is randomly drawn from a Normal distribution whose (mean −3 standard deviations) is 'a' and (mean +3 standard deviations) is 'b'. This procedure is then repeated to produce the set of data as required.

8.4 TIME-SERIES SMOOTHING

Data is often presented as a time-series as shown in the figure. The independent variable need not be time but a counter treating the data as a sequence of values. The fundamental processing is one of smoothing (Vollmann *et al.*, 1992). Smoothing reduces the variability of the raw data which enables any underlying trend to be more easily seen. The dotted line in the figure is an 11-point moving average of the raw data. The

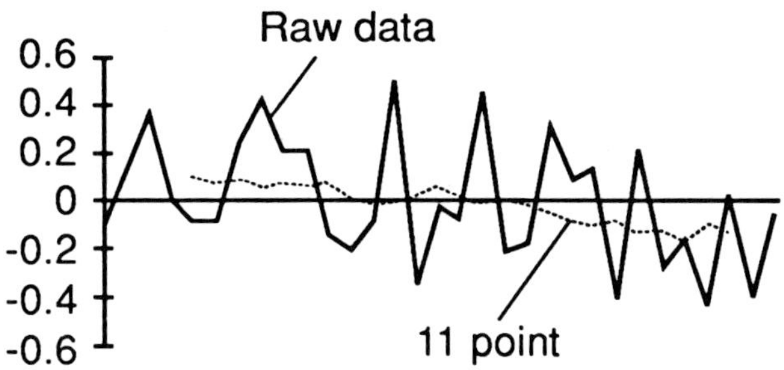

moving average is the simplest smoothing technique available. It is simply done by taking a fixed number of points, in this example 11, and calculating the average of these points. This value is placed at the mid-point of the sample. This is repeated by incrementing one data point and calculating the average of the next sample and so on. The amount of reduction in variability depends on the number of data points in the sample. The larger the sample the larger the reduction in variability.

An alternative method is called exponential smoothing, see figure

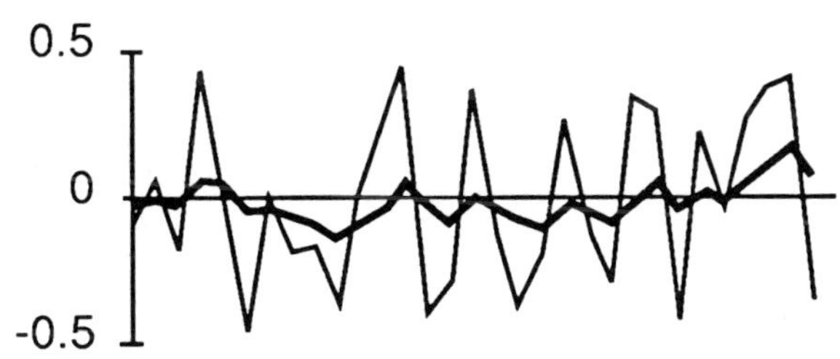

(Vollmann *et al.*, 1992). This is a recursive method as shown below:

$$E_{n+1} = E_n + \alpha[x_{n+1} - E_n], \text{ where}$$
E_{n+1} = current smoothed output
E_n = previous smoothed output
x_{n+1} = current raw data point
α = smoothing constant ($0 < \alpha < 1$).

A typical value for α is 0.2. This method has the advantage in that data is not 'lost' at the start and end of the data stream. Further, all the data is used with the data being discounted as it 'ages'. One can consider any smoothing scheme as being a weighted sum of the raw data. Consider a 3-point moving average. The scheme can be written as:

$$\text{output} = \frac{x_1 + x_2 + x_3}{3} = \frac{1}{3}x_1 + \frac{1}{3}x_2 + \frac{1}{3}x_3.$$

For a 3-point scheme the weights are $\frac{1}{3}$ and in general for a n-point scheme the weights are 1/n. The exponential scheme has successive weights that are $\alpha, \alpha(1-\alpha), \alpha(1-\alpha)^2, \alpha(1-\alpha)^3$, etc. In other words the weights of an exponential scheme form a geometric series whilst those of a moving average scheme are constant. Thus the moving average gives equal weight to the data points under consideration whilst the exponential gives less and less weight as the data 'ages'. In the example shown the moving average weights are 0.2 for the five points considered. The exponential scheme gives successive weights of 0.2, 0.16, 0.128 etc. to eventually all the data points.

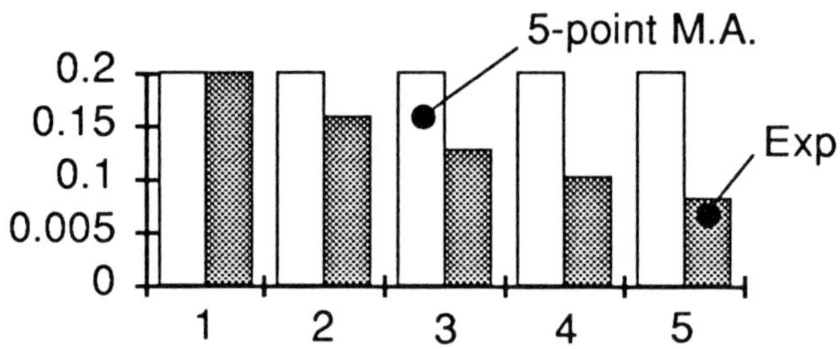

Weights of smoothing schemes

8.5 MODELS

We have already seen how a time-series may be used to indicate any trends by using a smoothing scheme. This is not the only procedure that can be undertaken. Often a law is fitted to the data usually using a least squares routine (Gerald and Wheatley, 1970). One must be clear why any law is fitted. Possibly there is a prior hypothesis what the law is and the data is then used to estimate the law's parameters. In fact this is the only valid use of the least squares technique. Alternatively, a convenient law is chosen, one that is convenient or tractable mathematically speaking. In

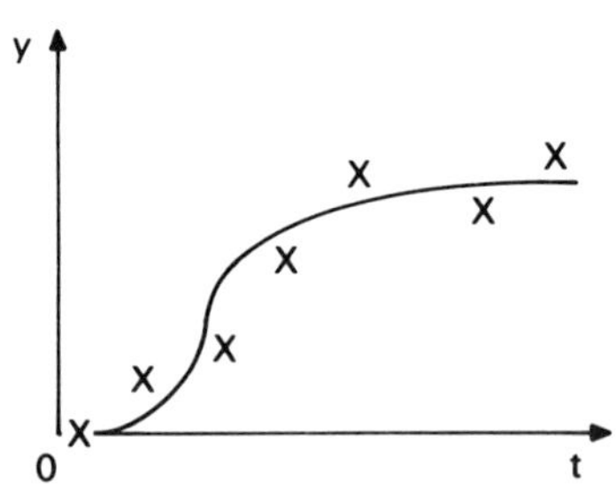

$y = f(t)$
minimize S
where $S = \sum_i [y_i - f(t_i)]^2$

this case the 'law' is not used as a description of the physical system but rather as a convenient way of representing the data. This use of a model enables the data to be stored simply by noting the parameters of the model. This method of representing data can be used as a source of data for simulation studies.

Using a model to generate typical data in a simulation has restrictions. It requires every generated data point to be adjusted to represent natural variation, for only using the model will repeatedly generate the same set of points. Further the model represents how the dependent variable varies against the independent variable. Far more useful is the generation of data from a given distribution.

8.6 RANKING DATA

In Chapter 7 we saw that the failure modes in CNC and NC machines were ranked in frequency of occurrence. This data was used to justify the selection of an axis drive as a pertinent sub-assembly to treat in a system manner. We now consider a method of optimizing the selection of a subset of items from a population that represent the dominant features of the population. The technique is due to Pareto (Oakland, 1987). It is best explained by way of an example. Consider some feature that is represented by a set of numbers, for instance a data set of 10 variables with particular variances (Table 8.1).

We wish to select a subset of variables that account for a significant amount of the total variance. We will return to this problem of total variance of a set of data later in this chapter. Suffice to assume at this time that the total variance can be represented by the sum of the individual variances. The aim is to select those variables that will account for say 80% of the total effect; in this case the total variance. The procedure is to rank the variances in descending order; then form a cumulative sum; finally calculate the percentage of the total at each intermediate cumulative sum. In the example in the following figure the first cumulative point is the first point, 9.7. The next cumulative sum is 9.7 plus the second point, 7.3, giving 17.0. This procedure is repeated until we reach the final total

Table 8.1 Data set of 10 variables with variances

Variable	*1*	*2*	*3*	*4*	*5*	*6*	*7*	*8*	*9*	*10*
Variance	1.2	3.6	7.3	0.9	2.3	5.3	4.8	9.7	6.2	5.8
Re-order	(8)– 9.7	(3)– 7.3	(9)– 6.2	(10)– 5.8	(6)– 5.3	(7)– 4.8	(2)– 3.6	(5)– 2.3	(1)– 1.2	(4)– 0.9
Cum. sum	9.7	17.0	23.2	29.0	34.3	39.1	42.7	45.0	46.2	47.1
Cum. %	20	36	49	62	73	83	91	96	98	100

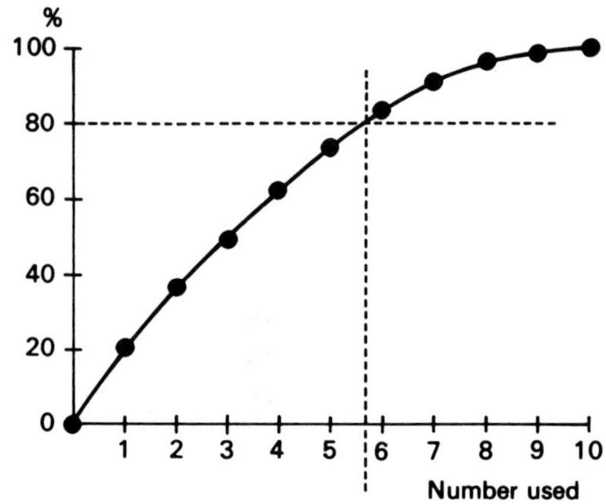

of 47.1. Then each intermediate point is written as a percentage of the total. Thus the first two percentages are (9.7* 100/47.1) giving 20% and (26.7* 100/47.1) giving 36%. Subsequently a convenient level is chosen, 80% say, and the number of variables needed to represent this percentage determined. In the example the 80% level falls between five and six variables. So five variables account for 73% of the total variance. Concentrating effort on these five variables (nos 8, 3, 9, 10 and 6) gives 73% cover.

8.7 PRINCIPAL COMPONENT ANALYSIS

The example in the Pareto analysis was selecting a significant percentage of the total variance of a set of data. CM data will consist of records of many variables of interest and the total variation of the set of records is a quantitative measure. One could argue that the total variance of the data is indicative of the health of the system generating the data. So if we were monitoring n parameters say, and collected m data points for each parameter the variances of the n parameter values could be estimated using the m data points per parameter value.

$$\mathrm{var}_k = \frac{\sum_{i=1}^{m} [x_{ik} - \mu_k]^2}{m-1}, \quad k = 1, 2 \ldots n.$$

Then the total variance is given by

$$\mathrm{var}_{tot} = \mathrm{var}_1 + \mathrm{var}_2 + \cdots + \mathrm{var}_n.$$

Unfortunately in general this is not true. As the parameters relate to one particular system it is likely that there will be a measure of 'correlation' between the measurements. The connection between two variables is measured by the covariance (Kraniauskas, 1992) of the corresponding two sets of data. Consider two parameters x and y for which m data points are available. Then:

$$\mathrm{cov}_{x,v} = \frac{\sum_{i=1}^{m} [(x_i - \mu_x)(v_i - \mu_v)]^2}{m - 1}$$

One can think of the variance as a covariance where the variables are the 'same variable'. It is an auto covariance of two variables. (This must not be confused with an auto covariance function (Kraniauskas, 1992) which calculates the covariance with successive shifts between the two data streams.) Thus the total variance of the data is:

$$\mathrm{var}_{tot} = \mathrm{var}_1 + \mathrm{var}_2 + \cdots + \mathrm{var}_n + \mathrm{cov}_{1,2} + \mathrm{cov}_{1,3} + \cdots + \mathrm{cov}_{2,1} + \cdots + \mathrm{cov}_{n-1,n}$$

It is more convenient to present this series as a matrix. Then the total variance is given by the covariance matrix, **C**.

$$\mathbf{C} = \begin{bmatrix} \mathrm{var}_{1,1} & \mathrm{cov}_{1,2} & \mathrm{cov}_{1,3} & \cdots\cdots & \mathrm{cov}_{1,n} \\ \mathrm{cov}_{2,1} & \mathrm{var}_{2,2} & \mathrm{cov}_{2,3} & \cdots\cdots & \mathrm{cov}_{2,n} \\ \mathrm{cov}_{3,1} & \mathrm{cov}_{3,2} & \mathrm{var}_{3,3} & \cdots\cdots & \mathrm{cov}_{3,n} \\ \vdots & \vdots & \vdots & & \vdots \\ \vdots & \vdots & \vdots & & \vdots \\ \mathrm{cov}_{n,1} & \mathrm{cov}_{n,2} & \mathrm{cov}_{n,3} & \cdots\cdots & \mathrm{var}_{n,n} \end{bmatrix}.$$

The leading diagonal has the variances of the measurements while the off-diagonal terms are the covariances. As the variances are the average of a squared quantity they are always positive. This is why one can always take the square root to obtain the standard deviation. The covariances can be positive or negative, indicating a positive or negative relationship between the two variables under consideration. It should be noticed that:

$$\mathrm{cov}_{1,2} = \mathrm{cov}_{2,1} \quad \text{or} \quad \text{in general } \mathrm{cov}_{i,k} = \mathrm{cov}_{k,i}.$$

In other words the covariance matrix is symmetrical. Further, all the entries are non-complex, which means that the matrix is square, real and symmetrical; useful characteristics for further processing.

The total variance of the measures is a vehicle for helping the management of process and manufacturing system improvement. Hand-in-hand with CM is quality in all its manifestations. Efficient pay-back in planning and maintenance effort is dependent on putting the effort that will give the greatest benefit. The Pareto analysis would give such information. However in this instance the total variance is not a sum of discrete measures but a matrix with coupling terms between the variables. What is needed is a decoupling technique that will enable the Pareto analysis to be implemented. The covariance matrix consists of entries that retain the units in which the various parameters were measured. Thus the range of numbers could be large. This is not inducive to robust numerical techniques; the matrix needs to be normalized. This is achieved as follows: Divide the i, j entry by

$$\sqrt{\text{var}_{i,i}\,\text{var}_{j,j}}.$$

This results in the entries being the simple correlation coefficients. Every element on the leading diagonal becomes 1. Thus the covariance matrix is normalized to give the correlation matrix:

$$\mathbf{R} = \begin{bmatrix} 1 & r_{1,2} & r_{1,3} & \cdots\cdots & r_{1,n} \\ r_{2,1} & 1 & r_{2,3} & \cdots\cdots & r_{2,n} \\ r_{3,1} & r_{3,2} & 1 & \cdots\cdots & r_{3,n} \\ \vdots & \vdots & \vdots & & \vdots \\ \vdots & \vdots & \vdots & & \vdots \\ r_{n,1} & r_{n,2} & r_{n,3} & \cdots\cdots & 1 \end{bmatrix}$$

Of course the measurements are still interlinked. The test data are a set of correlated data. The very nature of a system with the interdependence of various components and sub-assemblies naturally leads to data that exhibit this, sometimes weak, correlation. If there was no 'cross-talk' between the data the correlation matrix would simply have 1s on the leading diagonal and 0s elsewhere. Equivalently, the covariance matrix would have variances on the leading diagonal and all the covariances would be zero. In this situation we could say that the total variance is equal to the sum of the individual variances. This is the ideal. We need to decouple the variables so that we can concentrate on those contributing most to the total variance.

The technique is to **transform to orthogonal variables**. Orthogonal variables are **independent** variables. Knowledge of one variable gives no information of the other. The x–y axes of a rectangular coordinate system are examples of othogonal variables. The correlation matrix is the basis for a similarity transformation which de-couples the data. A geometrical example will illustrate the situation. Consider an ellipse centred at the origin with its axes coincident with coordinate axes and an ellipse that has been rotated and translated from the origin, A and B respectively. The equation for the translated ellipse is

$$B \Rightarrow ax^2 + by^2 + cxy + dx + ey + f = 0$$

and the ellipse at the origin

$$A \Rightarrow ax^2 + by^2 = c^2$$

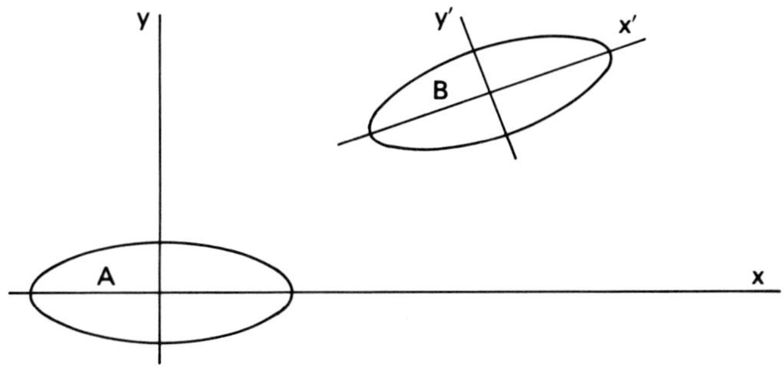

The equation for the ellipse at the origin does not have any 'xy' terms, the x and the y are 'independent'. Whilst the ellipse at the origin the x and the y are 'independent'. The minor and major axes of an ellipse are the principal axes of the ellipse. If the B ellipse is represented with respect to these principal axes (x', y') we get an equation in the form of A. This illustrates the equivalent data processing that is needed to decouple the test data. The technique uses the eigenvectors (and eigenvalues) of the correlation matrix (Kraniauskas, 1992). The fundamental relationship is given by:

$$\mathbf{Ru} = \lambda \mathbf{u}$$

where $\mathbf{R}$ = any square matrix (n × n) (in this application the correlation matrix)
$\mathbf{u}$ = column vector (n × 1)
λ = scalar.

The vector $\mathbf{u}$ is called an eigenvector and the scalar, λ, an eigenvalue. The eigenvector and associated eigenvalue are not unique. For a (n × n) matrix there will be n eigenvectors with corresponding eigenvalues. One can consider the eigenvector as a vector that when pre-multiplied by $\mathbf{R}$ produces a vector **in the same direction** but scaled by λ, the eigenvalue. The eigenvectors form an **orthogonal set**. If we transform our data using this set then we decouple the data.

The set of eigenvectors can be formed into a matrix:

$$\mathbf{u}_1 = \begin{bmatrix} u_{11} \\ u_{12} \\ u_{13} \\ . \\ . \\ u_{1n} \end{bmatrix} \quad \mathbf{u}_2 = \begin{bmatrix} u_{21} \\ u_{22} \\ u_{23} \\ . \\ . \\ u_{2n} \end{bmatrix} \quad \mathbf{u}_3 = \begin{bmatrix} u_{31} \\ u_{32} \\ u_{33} \\ . \\ . \\ u_{3n} \end{bmatrix} \cdots\cdots \mathbf{u}_n = \begin{bmatrix} u_{n1} \\ u_{n2} \\ u_{n3} \\ . \\ . \\ u_{nn} \end{bmatrix} \Rightarrow [\mathbf{u}_1\, \mathbf{u}_2\, \mathbf{u}_3 \ldots \mathbf{u}_n]$$

$$\mathbf{U} = \begin{bmatrix} u_{11} & u_{21} & u_{31} & \cdots & u_{n1} \\ u_{12} & u_{22} & u_{32} & \cdots & u_{n2} \\ u_{13} & u_{23} & u_{33} & \cdots & u_{n3} \\ \cdot & \cdot & \cdot & \cdots & \cdot \\ \cdot & \cdot & \cdot & \cdots & \cdot \\ u_{1n} & u_{2n} & u_{3n} & \cdots & u_{nn} \end{bmatrix}$$

We use the matrix **U** and its inverse to decouple or diagonalize the matrix **R**. The resulting matrix is a diagonal matrix with the eigenvalues, λs, on the leading diagonal.

$$\mathbf{U}^{-1}\mathbf{R}\mathbf{U} = \mathbf{diag}[\lambda]$$

$$\mathbf{diag}[\lambda] = \begin{bmatrix} \lambda_1 & 0 & 0 & . & 0 \\ 0 & \lambda_2 & 0 & . & 0 \\ 0 & 0 & \lambda_3 & . & 0 \\ . & . & . & . & 0 \\ 0 & 0 & 0 & . & \lambda_n \end{bmatrix}$$

The original measurements are transformed to new variables called the **principal components** of the data.

If $x_i, i = 1, 2 \ldots n$ are the original measurement set and $z_i, i = 1, 2 \ldots n$ are the principal components, then the linear relationship between the z's and x's is given by:

$$\begin{aligned} z_1 &= u_{11}x_1 + u_{12}x_2 + u_{13}x_3 + \cdots\cdots + u_{1n}x_n \\ z_2 &= u_{21}x_1 + u_{22}x_2 + u_{23}x_3 + \cdots\cdots + u_{2n}x_n \\ z_3 &= u_{31}x_1 + u_{32}x_2 + u_{33}x_3 + \cdots\cdots + u_{3n}x_n \\ &\quad . \quad . \quad . \quad . \quad . \quad . \\ z_n &= u_{n1}x_1 + u_{n2}x_2 + u_{n3}x_3 + \cdots\cdots + u_{nn}x_n \end{aligned}$$

where $u_k = [u_{k1}, u_{k2}, u_{k3}, \cdots\cdots\cdots u_{kn}]^T$ is the kth eigenvector of the correlation matrix. The principal components, z's, are uncorrelated and what's more their variances are given by the corresponding eigenvalues, λ's. In fact the covariance matrix of the principal components is a diagonal matrix with the λ's on the leading diagonal. Thus the total variance which was represented by the correlation matrix of the original measurements has been distributed over the new variables, the principal components, as the λ's.

x's(correlated) ⇒ Transform ⇒ z's(uncorrelated)

$$\begin{bmatrix} \sigma_{11}^2 & \sigma_{12}^2 & \sigma_{13}^2 & . & \sigma_{1n}^2 \\ \sigma_{21}^2 & \sigma_{22}^2 & \sigma_{23}^2 & . & \sigma_{2n}^2 \\ \sigma_{31}^2 & \sigma_{32}^2 & \sigma_{33}^2 & . & \sigma_{3n}^2 \\ . & . & . & . & . \\ \sigma_{n1}^2 & \sigma_{n2}^2 & \sigma_{n3}^2 & . & \sigma_{nn}^2 \end{bmatrix} \Rightarrow \text{Transform} \Rightarrow \begin{bmatrix} \lambda_1 & 0 & 0 & . & 0 \\ 0 & \lambda_2 & 0 & . & 0 \\ 0 & 0 & \lambda_3 & . & 0 \\ . & . & . & . & 0 \\ 0 & 0 & 0 & . & \lambda_n \end{bmatrix}$$

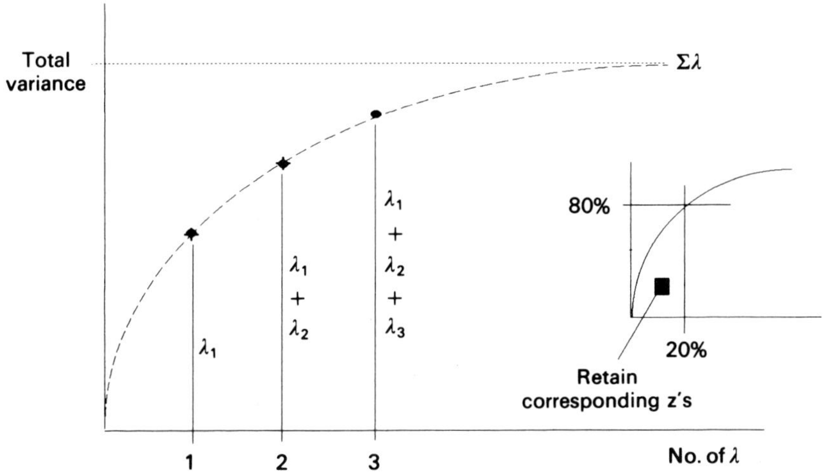

Thus we can now perform a Pareto analysis on the λ's. For convenience we assume that

$$\lambda_1 > \lambda_2 > \lambda_3 \cdots\cdots\cdots > \lambda_{n-1} > \lambda_n$$

The subset of the z's so chosen can be further analysed to see which of the original measurements (the x's) contribute most to each z. This is done by examining the weights represented by the u elements using a Pareto analysis on the elements making up the u's. This can be summed up diagramatically.

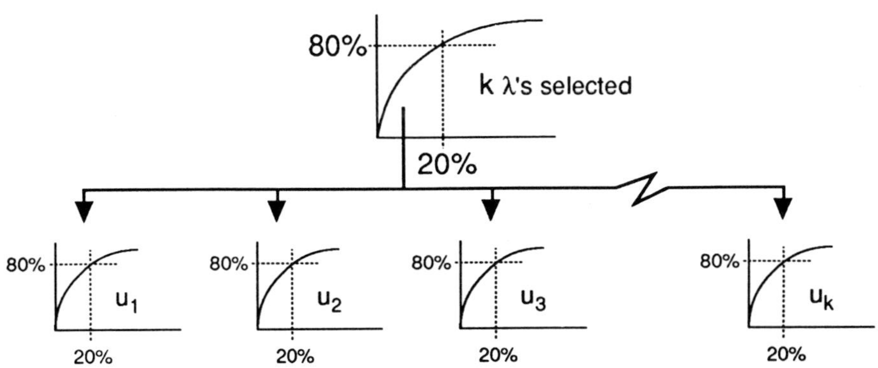

For example, suppose that z_1 was one of the significant principal components and that u_{11}, u_{16}, u_{18} and u_{19} were the dominant elements of the corresponding eigenvector. Then we can say that by monitoring z_1 we are in effect monitoring x_{11}, x_{16}, x_{18} and x_{19} collectively. The 80%/20% relationship is arbitrary, common sense should be used in selecting the

Table 8.2 Pareto analysis of eigenvalues ane resulting eigenvectors

Principal component	*Original measurements (x's)* 1	*2*	*3*	*4*	*6*	*7*	.....	*n−1*	*n*
Z1	X	X		X		X	.....		X
Z2		X	X				.....	X	
Z3	X			X	X		.....		X
....	.....	.....	.....	.....	.....	.....	.....	.....	.....
Z_k	X		X					X	X

cut-off point. In practice one never gets a point that lies exactly at 80%/20%. The higher the point the more cover we get but the cost is we process a higher number of variables. It should be stressed that the transformation produces variables, the principal components, that are in a different domain from reality. It is like the situation when we use the Laplace transform. To interpret the analysis one has to return to the original tests or measurements. One can summarize the procedure in tabular form. Table 8.2 summarizes the Pareto analysis of the eigenvalues and the resulting retained eigenvectors. It furnishes the link between the monitored principal components (z's) and the pertinent original measurements (x's). Further details can be found in Williams (1974).

8.8 HEALTH INDICES

We now consider a technique that has been developed specifically for CM (Hoh, 1992). The classification of the state of a system as either healthy or sick is a binary decision called checkout. Of critical importance is the

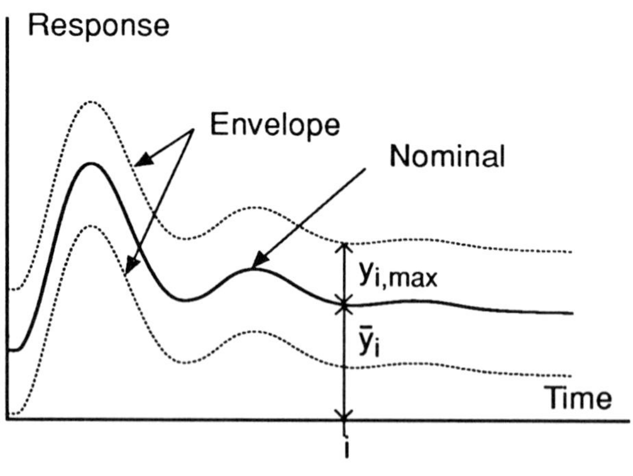

drift in performance from the healthy state to the sick state. In general all equipment will be 'off-colour' rather than completely 'fit'. Using a systems approach entails processing a dynamic response. In the past this response was displayed as a continuous curve and the skilled test engineer made his decision on the pattern or shape characteristics of the display. In automating this process one needs to sample the response and make decisions on a subset of response data. Of interest is the use of all the data over the dynamic range.

For the healthy system a test response is defined by n data points and has a value y_i at the ith point. If the test is repeated there will be a natural deviation, albeit small. Let m be the number of repeated tests and y_{ij} is the value of y_i for the jth test. The mean value of y_i is defined as:

$$\bar{y}_i = \frac{\sum_{j=1}^{m} y_{ij}}{m}$$

Thus the n values of $\bar{y}_i$ define the mean response which can be taken as the nominal response of the system. There is a need to define the spread of values about the mean. This is achieved by noting the maximum values of the y_i's for each ith point and taking these maximum values to define the envelope of healthy performance.

In the early days of diagnostics the minimum amount of data was stored and used in checkout and diagnostics. These days it is a trivial matter to obtain, store and process many hundreds of data points from a single transient response. However it is still a quest to obtain a single 'figure of merit' to describe the system's health. What would be useful is to condense the set of n measurements obtained in any single test or checkout into a single measure. There are various ways of doing this; here a 'quasi' standard deviation is used, called the Health Value (HV). This is defined for the jth test as:

$$HV_j = \sqrt{\frac{1}{n-1} \sum_{i=1}^{n} \left[\frac{y_{ij} - \bar{y}_i}{\bar{y}_i} \text{x} 100 \right]^2}$$

m values of $HV_j (j = 1, 2 \cdots m)$ are calculated. The largest of this set of values is called HV(nom),

$$\text{where } HV(\text{nom}) = \max(HV_j).$$

This then is the maximum variation over the m trials that occurs between an experimental response and the nominal response whilst the system is healthy.

For a possibly faulty system the response, over the same n points, will have values y_{if}, $i = 1, 2, \ldots n$. The same calculation is made yielding HV(f),

$$HV(f) = \sqrt{\frac{1}{n-1}\sum_{i=1}^{n}\left[\frac{y_{if} - \bar{y}_i}{\bar{y}_i} x100\right]^2}$$

HV(f) measures the average deviation of the possibly faulty response from the nominal. The Health Index is defined as:

$$HI = \frac{HV(nom)}{HV(f)}$$

The ratio compares the movement from nominal of the worst healthy response with the movement from nominal of the possible faulty response. The value of HI will normally be less than one but values greater than one are possible. This will occur when the response of the tested system 'lies among' the same range of the healthy system. Thus a calculation of HI yielding a value greater than one indicates a healthy system. We then set the condition that if:

$$\begin{aligned} HI \geqslant 1, &\quad \text{then } HI = 1 \\ HI < 1, &\quad \text{then } HI = HI \end{aligned}$$

There are a few precautions that have to be taken in using HI. There is the natural numeric problems near the origin where values in HV( ) are small. Also there can arise the condition where the faulty response moves outside the max/min spread only at a localized part of the response and thus the HI may not indicate any loss of health. This can easily be checked by the usual system checkout procedure which would be used first.

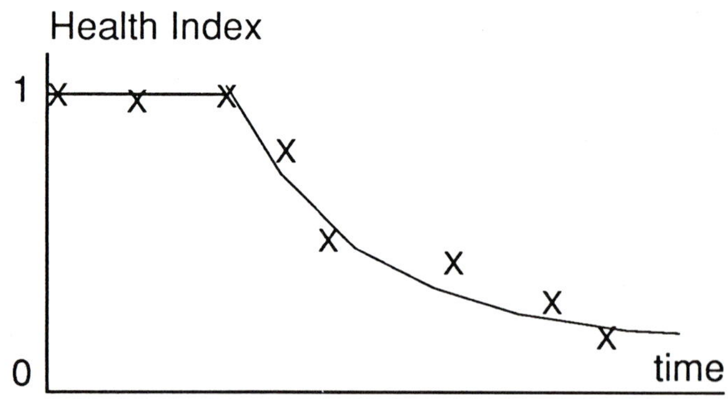

HI will vary with time. As wear and degradation takes place HI will fall below one. This drift can be monitored and gives a warning of future system or subsystem falling out of specification. This will enable efficient maintenance action to be planned. Further, the nature of the values that HI can take (between $0 \rightarrow 1$) makes it an ideal vehicle for plant wide monitoring. It means that the condition of a system under CM is normalized to a figure ranging between 0 and 1 regardless of the nature of the measurement or subsystem type. This is a prerequisite for an effective expert system based CM.

8.9 REFERENCES

Gerald, C.F. and Wheatley, P.O. (1970) *Applied Numerical Analysis*, Addison-Wesley.

Hoh, S.M. (1992) Condition Monitoring of Machine Tools. Ph.D. Thesis, University of Wales, Cardiff.

Johanson, N.L. and Leone, F.C. (1964) *Statistics and Experimental Design*, Vol. 1, John Wiley & Sons, New York.

Kraniauskas, P. (1992) *Transforms in Signals and Systems*, Addison-Wesley.

Oakland, J.S. (1987) *Statistical Process Control*, Heinemann, London.

Vollmann, T.E., Berry, W.L. and Whymark, D.C. (1992) *Manufacturing Planning and Control Systems*, Irwin, Boston.

Williams, J.H. (1974) Principal component analysis of production data. *IERE Trans.* **44**, 9, 473–488.

9 Future developments of condition-based maintenance in manufacture

9.1 INTRODUCTION

The introduction and use of Predictive Condition Monitoring (PCM) to enhance both machine and maintenance performance via the application of suitable information processing, analysis, and surveillance techniques, now appears to be rapidly gaining favour within manufacturing industry due to:

> The advent of powerful, accurate, reliable and non-contacting, 'intelligent' electronic sensing systems, which are affordable, and can be linked via computer/software technology, to provide the timely and cost effective control of advanced manufacturing equipment.

Developments in the field of Information Technology (IT) currently allow for 'real time' data acquisition and analysis to be performed during the production process, and consequently permit the useful implementation of many previously theoretical techniques in Condition Monitoring (CM). In turn, such technology enables the design of advanced manufacturing systems, which through the use of 'machine intelligence', attempt to achieve the ideal concept of unmanned production (Wright and Bourne, 1988).

There are several advantages to be obtained by actively pursuing the goal of implementing manufacturing/machine system intelligence, each of which is related to manufacturing economics and the pursuit of cost reduction. Accordingly they may be enumerated as follows (Williams and Davies, 1986):

1. A reduction in machining cycle times and hence cost, via the use of 'intelligent' adaptive control to achieve optimal machining conditions, in-process/in-cycle gauging, automatic workpiece/tool set-up to minimize second operation work, and the optimization of tool life during processing.
2. A reduction in the number of defective components and hence cost, via the use of 'intelligent' adaptive control for quality monitoring, in-process inspection, machine, system, coolant, and cutting tool surveillance.
3. A reduction in manufacturing lead time and hence cost, via the use of 'intelligent' adaptive control, both in-process and in-cycle to improve system availability for production, safety, reliability, diagnostics and maintainability.

In this context, the common factor required to achieve these benefits can readily be identified as 'intelligent' adaptive control, and this may be defined as:

> Providing a manufacturing/machining system with the reasoning capability and knowledge to adaptively control its actions and thereby optimize its operation in response to given criteria, environmental, and system stimuli.

In concept, such sensor-based machine tool management systems are not new, indeed, several experimental arrangements have been developed under the guise of adaptive control for machining in the past (Milner, 1975). However these systems have always been 'imperfect', due in the main to either insufficiently advanced technology, or 'incompleteness' in that although the technology worked, the 'whole machine' was not monitored or managed in an optimal way.

Now however, and within the context of machining processes, the state-of-the-art is such that the technology exists to implement this concept on computer numerical control (CNC) machine tools, and by implication, in flexible manufacturing systems or cells (FMS/FMC) (Wright and Bourne, 1988). Machine intelligence when combined with adaptive control and condition monitoring has the potential to increase equipment utilization, reduce labour costs, reduce inventory and floor space requirements, improve component quality and extend system life (Siefried, 1985).

Accordingly, this chapter will initially concentrate on outlining advances in this area of condition-based maintenance (CBM). Subsequently, the chapter will describe the implications for manufacturing industry when the economic arguments in favour of developing a self optimizing machining system are accepted, the technical problems are overcome, and this concept becomes a commonplace reality in a modern production environment.

9.2 INTELLIGENT SENSORS

As new manufacturing systems move towards full automation and achieve an unmanned capability, the efficient gathering and processing of information becomes an essential requirement of the control system and its associated software. Accurate real time knowledge of both the system and the process is necessary to optimize the overall performance of the machine and hence maximize utilization and productivity. It is essential to note that:

> These two basic information channels, i.e. one controlling the automated process and the other monitoring the health of the system, need to interact continuously in the machine controller, so that suitable modifying action may be taken when non-optimal or fault conditions are detected.

Data acquisition is therefore seen to be a key requirement in this type of technology, and the sensor systems necessary to obtain useful data are self-evidently critical elements in the makeup of such a machining system. In theory, much of the sensing aspect of this technology is already available; however improvements are needed in size, cost, capability, resilience and reliability (Towill, 1985).

Recent developments in integrated circuit technology have progressed to the point where the assimilation of various sensor functions in a single unit has been achieved. In addition, information from several sources may also be combined within the package to form what might be termed an 'intelligent' sensor. Figure 9.1 illustrates the concept behind such a sensor, which for implementation would rely heavily on very large scale integration (VLSI) technology. As shown in the diagram, 'intelligent' sensors may well contain some data evaluation capability, such as that required to improve the signal-to-noise ratio or combine several input signals. This quantification would be done via a microprocessor included within a microminiaturized and integrated package (Towill, 1985).

Accordingly it is important to note that a great deal of research work still has to be done into the potential characteristics of 'intelligent' sensors, the main areas of this effort being:

1. The precise circumstances under which the operation to transform the generated signal should be integrated within the sensor. This may only be justified if an output feature such as the autocorrelation function (AF), or Fast Fourier Transform (FFT) of a signal, is found to be useful, in for example a sensor-based machine tool management system, and where there is a cost or efficiency benefit to be achieved from 'local' processing (Towill, 1985).

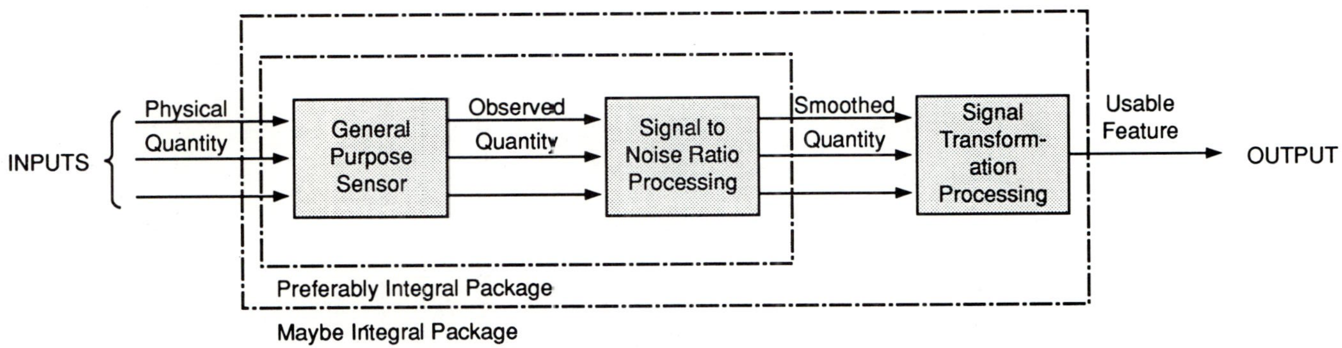

Fig. 9.1 Concept of a 'smart' sensor (Towill, 1985). N.B. Output feature might be pictorial information, surface description or a figure of merit e.g. Health Index etc.

2. The method by which the identification, selection and combination of input signals is organized into a meaningful quantity which constitutes the output of such an 'intelligent' sensor system. Typical multi-dimensional output features, as shown in Fig. 9.1, may be pictorial data, surface information or health index values depending on the application. VLSI miniaturization techniques now permit the flexible coupling of input information, so that various combinations of signal may be obtained with the goal of achieving, for example, indirect measurement (Ohba, 1992).

The development of new sensors and sensing systems is thus seen to be a very active area of engineering research. In many respects it is the key to effective condition-based maintenance, for as previously explained, without the ability to acquire accurate information, the quantitative monitoring of process machinery would be impossible.

Hence, the increasing use of computer-controlled production equipment in UK manufacturing industry, and the associated demand for its high productive availability, is now serving to emphasize the need for, and importance of, 'Real Time Actual Condition Knowledge' (RTACK). This information, the effective provision of which, can only be undertaken by 'intelligent' sensing systems, is now being correctly recognized as an absolutely essential requirement for the design of a true 'smart' machining system.

To develop such a self-optimizing machining system, RTACK information is required in several areas of this type of plant, and because the operating environment is difficult, these sensing systems need to be (Grosvenor and Martin, 1985):

1. fast, accurate and self-correcting in operation,
2. simple in design and rugged in construction,
3. non-intrusive and preferably non-contacting,
4. highly reliable, inducing no increase in system complexity.

For machine tool systems the areas of interest with respect to the development of intelligent sensor technology are (Davies, 1990) as follows.

Tool wear and breakage: This area covers firstly, the gradual wear of the tool during normal cutting conditions which is a progressive parameter, and secondly, the sudden cracking, chipping or fracture of the tool which results in catastrophic failure. Many of the sensing devices which are available for tool wear are also suitable for detecting tool failure, the main difference being in the signal analysis techniques applied to the data emanating from the sensors. In-process, in-cycle and out-of-cycle techniques are available as monitoring systems, using direct and indirect measurement. Table 9.1 outlines some of the many techniques currently used or under development in this area of sensing technology.

Table 9.1 Main areas of interest in the development of intelligent sensor systems for machine tools

1. *Tool wear and breakage:*
- Tool life modelling/monitoring by sensing machining time or the number of components manufactured.
- Examination of the tool edge for cracking by pneumatic or optical sensing.
- Tool size checking using touch trigger probes, optical sensors, proximity switches, ultrasonics, air and feeler gauging.
- Tool wear monitoring using radioactivity, analysis of wear particles, electrical resistance or current sensing.
- Optical methods of wear assessment using reflection, fibre optics, closed circuit television cameras and laser speckle pattern interferometry.
- Tool wear surveillance using temperature measurements obtained via thermocouple or infra-red sensing.
- Tool wear and breakage inspection using power consumption, motor current, bearing force, tool force, shaft torque, vibration, surface finish, workpiece dimensions, acoustic emission and ballscrew monitoring.

2. *Component size, surface finish, alignment, position and location:*
- Dimensional measurement of components via optical methods, touch trigger probes, laser, ultrasonic, pneumatic, capacitive and inductive sensing. Eddy current, microwave and electromechanical systems, in-line coordinate measuring machines and programmable automatic gauging.
- Surface finish measurement of components via stylus, interferometry, geometrical optics, holography, laser methods, differential AC interferometry. Optical Fourier Transform, scattering methods, ellipsometry, speckle techniques, image analysis and capacitance.
- Alignment, position and location of components via touch trigger probes, laser systems, moire fringe gratings, electronic speckle techniques and pneumatic methods. Eddy current proximity sensors, capacitive/inductive sensors, optical proximity sensors and vision systems.

3. *The machining system:*
- Force monitoring, vibration analysis, dynamic response testing, thermal monitoring, input/output analysis, accuracy trend detection, fluid flow sensing, motor current and wear monitoring.

Component size, surface finish, alignment, position and location: The measurement of the dimensions of the component being machined is a controversial area in monitoring systems. Much of the discussion revolves around whether this is best done in the machine tool, at an in-cycle measuring station or at an out-of-cycle inspection station. Very little work has been reported in which true in-process measurements have been made during machining, although as shown in Table 9.1 research in this area is proceeding. The true in-process measurement of surface finish for example is at present regarded as impossible or too costly to contemplate; however as shown in Table 9.1 this has not dissuaded research into its possible achievement. Similarly a fundamental requirement of unmanned machining is the alignment, positioning and location of components; within this area, again as shown in Table 9.1, several techniques are available or are under development.

The machining system: Several sectors within the machine itself require monitoring during the cutting process to effect unmanned operation, and these include:

1. The structure for vibration and/or thermal distortion.
2. The fluid systems for quality, levels, flows and leaks.
3. The control system for software, temperature and electrical or electronic failure.
4. The drive systems for performance and wear factors, such as backlash, dynamic response and friction, together with servo elements, such as motors, tachos, position transducers, transmission arrangements and couplings.

Bearings in particular need monitoring for damage as do other components of the machine such as relays or switches etc. Table 9.1 also depicts the techniques currently available or under development in this area.

The above outline gives a brief indication of the range of factors involved in the development of intelligent sensors for machining systems, and the sensing methods/requirements which now exist, or are under investigation for the monitoring of this type of production system. Useful signal acquisition is, however, only the first step in the creation of a practical unmanned machining environment. For, as with a human body, the RTACK information provided by the sensor system requires both interpretation and understanding before an intelligent response can be provided. Thus we need to consider both the problems and the developments associated with the identification and subsequent management of the RTACK data necessary to accomplish intelligent unmanned machining.

9.3 KNOWLEDGE REQUIREMENT

The current trend towards automated production systems presages the removal of all human supervision within a manufacturing environment, and this, in the case of machining systems, requires the replacement of 'operator' real time actual condition knowledge with its 'shadow' of sensor acquired RTACK data. In turn, the use of RTACK information to 'intelligently' control a machining system requires the investigation and specification of practical solutions to problems in the following areas (Ohba, 1992):

1. Multi-dimensional sensor processing: This in essence is the trade-off between utilizing all individual sensor information, separately, in the

control of a system, or its reduction, and subsequent collation prior to use, which results in control being effected via a data condensate.

2. Fused sensor processing: Essentially the use of sensors in combination, either utilizing series and parallel configurations, or through data 'overlap' to acquire 'new' control information.
3. Hierarchical sensor control systems: The multi-layering of sensor input at various levels in the control system to achieve optimum performance.

In the case of machining systems the last named area is worthy of further

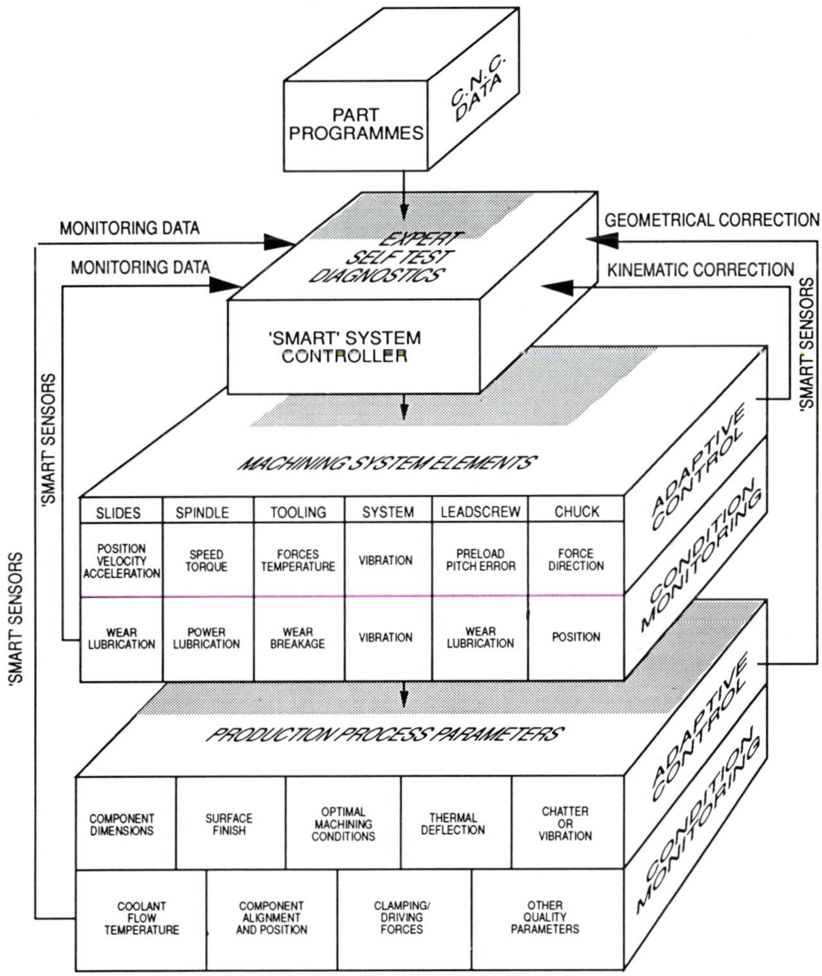

Fig. 9.2 The range of typical monitoring tasks required in an intelligent machine tool system.

study, firstly as an example of hierarchical control, and further to illustrate the use of multi-layer sensor information in unmanned production. Figure 9.2 outlines in conceptual form the monitoring and control tasks required of an intelligent machining system, highlighting the process and machine parameters which need to be measured, if possible, during the cutting operation. A key element in the concept is that the resulting information, essentially RTACK data, is used within an intelligent adaptive control system to optimize machine operation in total, and not simply to achieve sub-optimal goals such as efficient machining conditions.

Table 9.2 Typical monitoring tasks required of an Intelligent Machining System and relating to the machine itself

Machine component	*Monitored parameter*	*In-process real time*	*Adaptive control*	*Condition monitoring*	*Programme requirement*
Slides	Displacement	yes	yes	yes	yes
	Velocity	yes	yes	yes	yes
	Acceleration	yes	yes	yes	yes
	Lubrication	no	no	yes	no
Servo motors	Current	yes	no	yes	no
	Power	yes	no	yes	no
Spindle	Speed	yes	yes	yes	yes
	Forces	yes	yes	yes	no
	Torque	yes	yes	yes	no
	Position	no	no	no	yes
Spindle motor	Current	yes	no	yes	no
	Power	yes	no	yes	no
Cutting tool	Number	no	no	no	yes
	Forces	yes	yes	yes	no
	Temperature	yes	yes	yes	no
	Wear	yes	yes	yes	yes
	Breakage	yes	yes	yes	yes
Machine system	Vibration	yes	yes	yes	no
	Temperature	yes	yes	yes	no
Lead-screws	Preload	no	no	yes	no
	Lubrication	no	no	yes	no
Chuck	Force	no	no	yes	yes
	Direction	no	no	no	yes
	Position	no	no	no	yes
Control system	Temperature	no	no	yes	no
Coolant system	On/off	no	no	no	yes
	Flowrate	no	no	yes	no
	Temperature	no	no	yes	no
Oil system	On/off	no	no	no	yes
	Pressure	no	no	yes	no

In this context the techniques of system sensing, condition monitoring, fault diagnosis, quality surveillance, knowledge-based systems, neural nets, adaptive and autonomous control all need to be considered along with process and tool monitoring when formulating the design of an intelligent machining system. In particular the design of the system controller is crucial, for in addition to manipulating the speed, position, velocity and acceleration of the machine spindle and slides it must now interpret and command the operation of several ancillary systems on the machine, including by definition an integrated sensing system of some complexity.

Tables 9.2 and 9.3 contain the range of perceived monitoring tasks required of an intelligent machining system, and as shown, many of these functions need to be conducted in 'real time'. Thus a requirement for devolved parallel processing via 'smart' sensors, transputers and hierarchical control systems is an obvious outcome of pursuing this design. The processing requirement for each data channel is a function of the speed of usage of the information within the control system. It is also observable from Tables 9.2 and 9.3 that many of the monitoring tasks need to be performed, as previously stated, during the machining process. This reinforces the need already identified for sensors which can survive in hostile environments, and yet provide accurate, refined information about the operation in progress.

Unfortunately, there are few currently available sensor systems which have the capability to provide RTACK data during the actual production

Table 9.3 Typical monitoring tasks required of an intelligent machining system and relating to the component being machined

Workpiece feature	*Monitored parameter*	*In-process real time*	*Adaptive control*	*Condition monitoring*	*Programme requirement*
Component dimensions	Size	yes	yes	yes	yes
Surface finish	Value	yes	yes	yes	no
Optimal machining conditions	Index	yes	yes	yes	no
Thermal deflection	Size	yes	yes	yes	no
Component position	Datum	no	no	no	yes
Component alignment	Datum	yes	yes	no	no
Component clamping	Force	no	no	no	yes

process, and many are regarded as being too unreliable or ineffective for use during unmanned machining. In the area of tool wear or failure for example, extensive research is underway using force, power, vibration, acoustic emission and other system sensing techniques to find a suitable monitoring method. However many of these devices or methods have yet to show that they can distinguish clearly between wear and breakage of the tool, and to do so in a repeatable manner (Grosvenor and Martin, 1985).

In the case of surface finish, progress is being made via non-contact optical sensors, although overcoming the difficulties associated with heat, swarf, coolant, and smoke generation during the machining operation has yet to be satisfactorily solved. For in-process alignment, positioning and precise location of the workpiece, vision or fast three-dimensional laser scanning systems appear to offer the best solution, if these can be developed, to provide accurate and repeatable measurements under the conditions of unmanned machining (Grosvenor and Martin, 1985).

Great benefits await the development of true in-process sensing systems, including the elimination of in-cycle measurement and component washing stations. Each of these represents additional cost in terms of capital outlay, floor space occupied, control complexity, system maintenance, workpart transportation and extended production cycle times. Thus the economics of unmanned production are firmly in favour of developing in-process sensing, which as previously outlined, provides an elegant solution to the provision of RTACK data, and thus is vital for the creation of autonomous machining systems (Williams and Davies, 1986).

9.4 SYSTEM MANAGEMENT

The formulation of 'smart system' software to control the machine and provide its 'intelligence' appears at first glance to be a reasonably straightforward task, assuming that the appropriate sensor technology is available, along with the necessary electronic hardware. However when examined in detail, a more complicated situation is revealed, the resolution of which, it is suspected, would involve the use of artificial intelligence concepts, neural nets and in particular the use of expert knowledge-based systems (EKBS). Figure 9.2 illustrates the complex sensor hierarchy involved in the design of such a system, to which must be added the adaptive criteria for optimizing both process and machine plus the reasoning power to 'learn', resolve conflict, limit instability, and bestow autonomous 'intelligent' control.

Adaptive Control (AC) may be defined as:

> A technique that allows a system equipped with sensors to detect a changing environment, and if the change is detrimental, to automatically institute corrective action, such that the system being controlled is optimized in line with a given criterion.

Within this definition, current machine tool adaptive control systems can be classified into two major subgroups, these are Technological Adaptive Control (TAC) and Geometrical Adaptive Control (GAC). The first of these subgroups (TAC) contains two distinct system types, these are Adaptive Control of Constraints (ACC) and Adaptive Control of Optimization (ACO). Each method is formulated around a distinct principle, which is embodied in the particular technique definition as outlined below:

1. Adaptive control of constraints is a technique which focuses on achieving safe system operation under certain machine related physical constraints.
2. Adaptive control of optimization is a technique which focuses on the optimal achievement of economic production criteria.

Both methods are geared to a manufacturing environment, and assess optimality within the feedrate/cutting speed dimension, with adaptive control being effected via the adjustment of spindle speed or slide movement. In contrast, geometrical adaptive control (GAC) is product oriented and may be defined as follows:

3. Geometrical adaptive control seeks to optimize component conformity to a specification, via adjustments in tool position, and thereby to compensate for geometric deviations which may occur due to, for example, tool wear during the cutting process.

It is now easy to see why, given the foregoing discussion, the development of sensors and monitoring techniques in the areas of tool wear, breakage, surface finish, component setup and machine health are so important (Wright and Bourne, 1988). For without accurate and reliable RTACK data, learning would not take place, and efficient intelligent adaptive control would be impossible to implement.

Adaptive control of constraints is the most well developed of the three techniques, having been applied to both turning and grinding systems (Milner, 1975), while adaptive control of optimization and geometrical adaptive control are still the subject of research with rare commercial applications. However as computing and sensor power increases, both techniques may become viable, through the development of intelligent machining systems, which conceivably could use a combination of these three methods as part of its control arrangement (Williams and Davies,

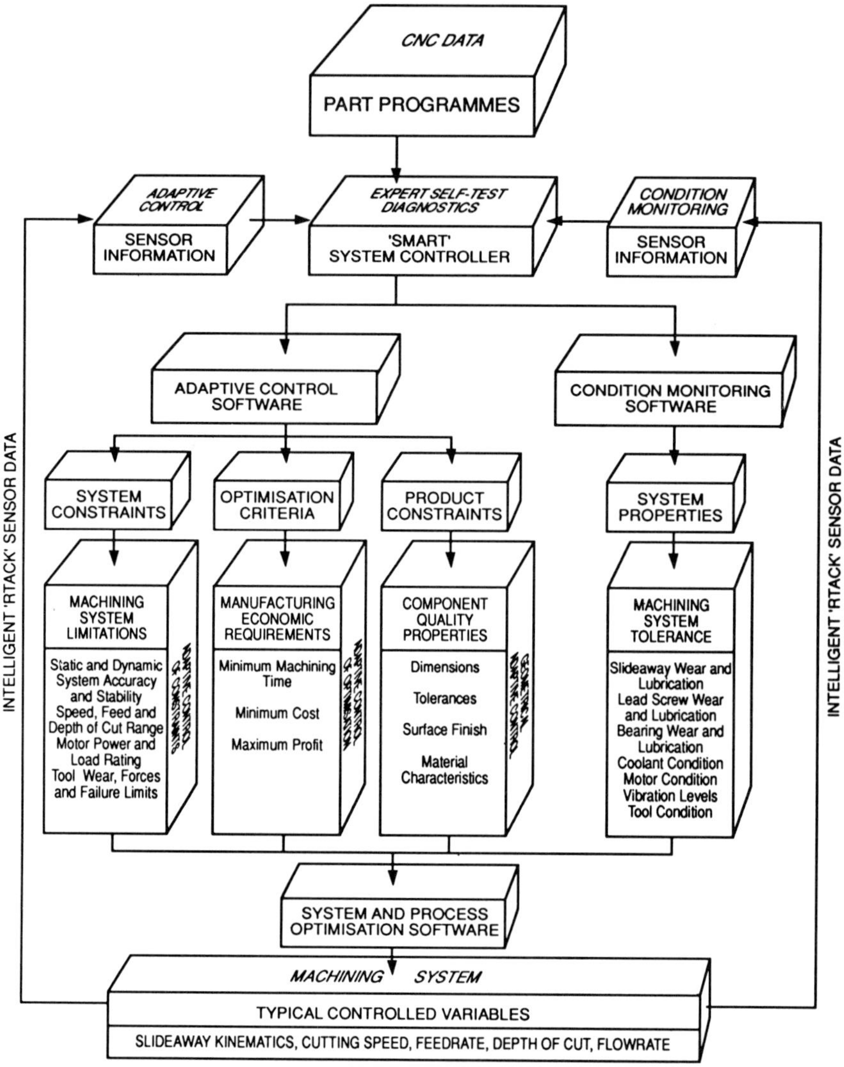

Fig. 9.3 The interaction of sensor information in a 'smart' manufacturing system.

1986). This integrated approach is illustrated in Fig. 9.3 which incorporates each method of adaptive control, together with the use of condition monitoring, to form the basis of a 'smart' machining system.

As intimated previously, a hierarchical control system is needed to manage the sensor/software arrangement required to make this concept a reality, and some work has already been done in this area to assess concept feasibility. The use of intelligent sensors, existing system sensing, condition monitoring methods, machine diagnostics, knowledge-based software and AI techniques have all received attention in recent years. For implementation, a new generation of machine controller may be required; one in which RTACK information supplied by intelligent sensors and evaluated through adaptive control and condition monitoring techniques is presented for action to a reasoning autonomous knowledge-based system. This may well utilize a neural network arrangement, if reasoning speed and reliability problems can be overcome, and thereby provide the intelligence required for a machine to achieve self optimization.

9.5 SMART SYSTEMS

In operation, a smart machining system would require a controller which retains the existing functional processing capability of current CNC machine tools for executing part programmes, issuing system commands and monitoring position feedback. In addition, it would require as previously outlined, a hierarchical parallel processing capability to achieve both real time adaptive control and the condition monitoring of machine plus the cutting process. Figure 9.4 illustrates in conceptual form the main requirements for such a system, with the twin problems of data processing time and system instability being overcome by the use of artificial intelligence software, parallel processing hardware, and the integration of intelligent active and passive sensor networks (Williams and Davies, 1986).

Within this arrangement, the parallel processors are free to monitor their part of the sensor network in an independent and dedicated manner, only interacting with the functional processor when a fault or disturbance is identified and action is required. Thus the functional system reliability should remain relatively unaffected by sensor system complexity, provided adequate safeguards are incorporated to 'lock out' sensor system failures and the intelligent sensor networks have a 'built-in' self-test capability. In this arrangement instability can be reduced by known faults being weighted for relevance to a given machining operation. Thus when a fault is detected, the control system is allowed discretion within the reasoning software to ignore deviations from nominal machining conditions which are not immediately detrimental to either process or machine. This integration of sensor, monitoring and software techniques makes available a known environment within which the controller can choose to act via the functional processor or merely report an error for maintenance action.

It is envisaged that the machining system failures or disturbances can be classified via a two-part signature which relates to active and passive sensor signals. Thus when the system is operating normally or in a

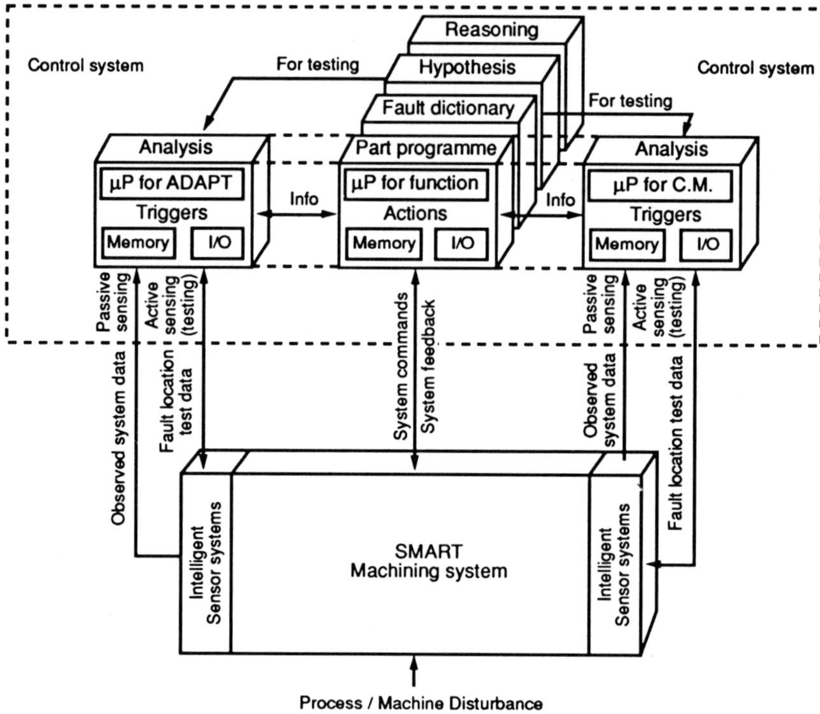

Fig. 9.4 A conceptual design for the intelligent control of a machining system. μP = Microprocessor. ADAPT = Adaptive control. C.M. = Condition Monitoring I/O = Input/Output.

'healthy' condition, only the passive or functional RTACK system sensor data would be monitored, thereby saving processing time and ensuring a fast response. These signals could possibly provide complete fault identification as outlined previously, or more likely the first indication of error together with part of the fault signature which on detection would initiate the checking of the fault dictionary for possible causes.

Alternative hypotheses may then be checked by the expert diagnostic system eliciting additional information via the active sensors using test signal injection or standard acquisition techniques. The RTACK data obtained would form the second part of the fault signature, and pattern matching of the total signature thus acquired would provide the controller with fault identification via the fault dictionary. Only when a fault is confirmed and corrective action required, would this be signalled to the functional processor for adaptive implementation (Gopal, 1991).

The logic of the surveillance concept outlined above can be illustrated by considering a hypothetical machining situation. Within the control system the query may arise as to whether the material is being cut at an optimum rate, the optimum being achieved when the predicted values for various criteria such as spindle torque, table feedrate and tool forces are actually present on the system during machining. Note that these typical parameters can be balanced by weighted importance to provide an optimum signature, and that tolerances may be added to each value thereby making provision for non-nominal conditions.

The system's motivation for seeking the optimum would be provided by a range of weighted goal seeking strategies, for machining rate may be only one of several optimum signatures monitored by the system during machining. The knowledge base would comprise relevant rules developed from applied science. These, in conjunction with a continuously updated RTACK database (provided by the sensor system), would be the mechanism whereby the current parameter values are determined and assessed against the optimal signature.

Should, for example, a machining rate parameter move out of tolerance, the passive sensors in either the functional or adaptive control system will detect it and the controller will seek to identify the error by the pattern matching technique outlined above. Essentially the controller is searching for additional knowledge relevant to the situation via the active sensors and its interaction with the machine monitoring network. The feedrate may be lower than that desired due to increased slideway friction. This is a condition which could be identified through 'passive' motor current sensing. Action may or may not be taken by the controller depending on the situation. The parameter error could be adjusted to nominal by an increase in the speed of the table motor, a shot of lubricant onto the slides or ignored.

The useful features of this arrangement are that, by the use of programmable and adjustable weightings and tolerances, non-critical deviation may be ignored while trends can be recorded and evaluated to predict future problems and alert maintenance services. The reactive/predictive nature of adaptive control and condition monitoring as proposed above, provides for the forward and backward chaining typical of expert system software. The example shows how forward chaining can be useful through an adaptive control network, and backward chaining could also be achieved by the condition monitoring sensors recording an apparently unrelated event. Assuming this to be, for example, an environmental temperature rise, the software, using this as a trigger, could link it to, say, dimensional errors on the component arising from thermal drift in the machine tool.

By designing the system software to incorporate 'learning' techniques, the machine could become familiar with its own idiosyncratic nature and accordingly decide on error detection to ignore, compensate, shut down safely or request maintenance action. One interesting feature displayed by artificial intelligence software as embodied in neural nets and expert systems is their ability to 'reason' under uncertainty, a skill which is a prime requirement in unmanned machining. This ability when married to the attributes of adaptive control, condition monitoring and intelligent sensing forms the basis of the smart system concept. Additional improvements may also accrue within the overall manufacturing system from the use of unmanned machining and these will be discussed in the next section.

9.6 INTELLIGENT MANUFACTURING

The foregoing discussion has centred around the design of an unmanned intelligent machining system and the roles played in that concept by:

1. Advanced, possibly intelligent sensors, which incorporate VLSI technology, AI software and potentially, some new physical methods of measurement.
2. Condition monitoring and adaptive control techniques, implemented via expert system algorithms, parallel processing hardware and the use of RTACK data.
3. Autonomous reasoning within the control system, which may, conceivably, utilize neural networking as a fundamental component.

Hopefully, the difficulties, avenues of research, and current position in respect of unmanned machining, have been explained with reasonable clarity, so that the reader may now appreciate that, with all its detailed problems, a smart machine tool is only one element of an autonomous manufacturing system. With this in mind, two challenges have to be overcome before intelligent manufacture is possible (Wright and Bourne, 1988):

1. The integration of machine actions into a cohesive flexible unit which can thereby form the basis of a viable manufacturing sub-element. Such intelligent machine units are essential to the make up of a smart production system.
2. The elimination of the dependency on human expertise for setup, monitoring, evaluation, decision making and corrective action, with all the corresponding economic benefits this implies if it can be achieved.

The key to overcoming both these difficulties lies with the development of efficient communication techniques. Firstly within the machine itself, to link sensors with supervisory systems and thereby obtain access to the monitoring, control and reasoning software necessary to impart intelligence, and secondly within the wider context of the manufacturing system.

This last point involves the communication necessary between individual machine units which make up a manufacturing system, the overall controller responsible for integrating those machines, thereby forming the system, and the management information software which supports the system to produce a viable operation. In this latter case, many more difficulties have to be overcome before true intelligent manufacturing systems are created; these include:

1. The so called 'standalone' syndrome, where communication between machines of different manufacture, each of which may have differing control systems, is impossible, and thus all are unable to communicate with one another or with an overall controller. Developments in the General Motors Manufacturing Automation Protocol (MAP), and the Department of Trade and Industry's (DTI) 'Open Systems' initiative, are designed to overcome this problem.
2. The efficient design of communication systems to achieve intelligent control, for many existing machines have anarchical control, suitable for stand alone operation but difficult to integrate into intelligent systems. Accordingly, research into the use of both hierarchical and heterarchical control of manufacturing systems is ongoing with a view to resolving this problem.
3. The integration of system support software, which may include management information, maintenance diagnostics, system scheduling programmes etc., with the main control arrangement. In effect, designing 'whole' system software, which not only controls the machinery but also provides autonomous intelligence for the manufacturing system. To assist in achieving this goal, new and improved computer software tools and languages are continually being developed, a good example of which is the Cell Management Language (CML) (Bourne, 1986).
4. The development of true intelligent adaptive control, which incorporates the ability to learn and reason with regard to its environment, and to apply this ability in the control of a manufacturing system. Research and development in the areas of artificial intelligence, neural networking and knowledge-based systems will hopefully provide a resolution to problems in this area.

Encapsulated within these general areas are a mass of detailed difficulties which need to be overcome before intelligent manufacturing systems can be designed with any degree of certainty regarding their successful utilization. The problems range through the evolution of sensors, software, controls, machine and process mechanics to vision, robotics and machining expertise.

Not all will be solved easily, or in the near future: one thing is certain however, extensive research is underway to solve these problems and this will continue as long as the economic imperative provides a driving force (Wright and Bourne, 1988).

9.7 CONCLUSIONS

The foregoing has attempted to illustrate the role envisaged for condition monitoring within an automated manufacturing environment. Hopefully it has outlined how condition-based manufacturing systems are likely to develop in the future, and how condition monitoring will play a much wider role in that scenario than simply acting as a maintenance tool within this environment. Condition-based monitoring and control is an essential ingredient for the development of intelligent manufacturing systems, and as such it should be regarded in this light rather than that of a sophisticated maintenance technique.

9.8 REFERENCES

Bourne, D.A. (1986) CML: A meta-interpreter for manufacturing. *AI Magazine*, **7** (4), 86–96.

Davies, A. (1990) *Management Guide to Condition Monitoring in Manufacture*, Institution of Production Engineers, London.

Gopal, D.P.G. (1992) An expert system for machine tool breakdown diagnostics using the Kappa Shell. UWCC (ELSYM) MSc Thesis, Cardiff.

Grosvenor, R. and Martin, K.F. (1985) Review of sensors for machine tool application. *UWIST Technote EP 141*, Cardiff.

Milner, D.A. (1975) An introduction to adaptive control. *The Production Engineer*, **54** (3), 175–180.

Ohba, R. (1992) *Intelligent Sensor Technology*, John Wiley & Sons, Chichester.

Siefried, R. (1985) Unattended machining. *SME Technical Paper MR85-468*, Detroit.

Towill, D.R. (1985) Sensor based machine tool management systems. *UWIST Technote EP 110*, Cardiff.

Williams, J.H. and Davies, A. (1986) Controller based manufacturing system management – how close to reality? *The International Journal of Advanced Manufacturing Technology*, **1** (5), 61–74.

Wright, P.K. and Bourne, D.A. (1988) *Manufacturing Intelligence*, Addison-Wesley, Reading, MA.

Index